Portmadoc shortly before the First World War. The Harbour lies off to the right of the picture and the approach to Britannia Bridge can be seen on the extreme right. Festiniog tracks can be seen curving through the lower centre of the image as they pass behind the Welsh Slate Company's wharf area and the rear of the sheds that dominated the Oakeley Wharf. The Croesor Tramway is clearly visible as it crosses the High Street and runs to the left of the Festiniog tracks between the slate wharves and the wood yard. Just above the lower edge of the image the Croesor metals made a junction with the Festiniog - the 'zero-point' of the Croesor Tramway.

The Croesor Valley lies between the distinctly shaped Cnicht ("The Welsh Matterhorn") and the rising slopes of the Moelwyns.

Commercial Postcard via Martin Pritchard

Copyright © 2018

by Dave Southern and Welsh Highland Railway Heritage Group

All rights reserved.

This book or any portion thereof may not be reproduced or used in any manner whatsoever without the express written permission of the publisher except for the use of brief quotations in a book review.

Printed in the United Kingdom

First edition 2018

ISBN 978-0-9930821-6-0

Printed by:

    ZPQ Designs

    Lytham St Annes

    FY8 1RD

Published by the Welsh Highland Railway Heritage Group

www.welshhighlandheritage.co.uk

Trade & Retail Sales:

    Welsh Highland Heritage Sales,

    c/o Adrian Gray

    25 The Pound

    Syresham

    BRACKLEY

    NN13 5HG.

    Email: sales@welshhighlandheritage.co.uk

Contents

		Page
Foreword		5
Preface		6
Acknowledgements		7
Miscellanea		8
Chapter 1.	Background	11
Chapter 2.	History of the Croesor Tramway	15
	Birth and Development	15
	Declining years and closure	19
	The final workings on the Croesor Tramway section under WHR ownership	20
Chapter 3.	The Route described	23
	Portmadoc to Gareg-hylldrem	23
	From the Inclines to Croesor village	30
	Cwm Croesor	32
	North of Blaen-y-Cwm	35
	Post-closure	36
Chapter 4.	Operations	37
Chapter 5.	Sources of Traffic	45
	Background	45
	The Quarries	46
	Other locations served by the Croesor Tramway	54
Chapter 6.	Personalities	59
Chapter 7.	Memories	67
Chapter 8.	Exploring what remains today	75
	From Porthmadog to Croesor Junction	75
	From Croesor Junction to Croesor Village	75
	A walk along the tramway from Croesor village	81
	Croesor Quarry	85
	Ceunant Parc	90
Chapter 9.	Welsh Highland Railway	93
Chapter 10.	The Tramway in use today	97
Appendix 1	A Specimen Waybill Summary Sheet	99
Appendix 2	Transcription of Moses Kellow's Letters	101
Appendix 3	Croesor Quarry - Output and Employment 1895 - 1941	113
Bibliography		115
Sources		116

The Croesor Tramway between the Upper Parc Incline and Croesor Village in the 1920s. At the end of the right-hand curve the track turns to the left and disappears behind the trees that mark the course of the Afon Croesor. The track crosses that river over a clapper bridge as it runs into the village. The village buildings are hidden by the trees.

The prominent building seen to the left of the second pole is Croesor-bâch. This small farmstead sits part way between Bryn-hyfryd and Croesor-fawr and on the opposite, left-hand, side of the tramway.

The striking peak of Cnicht dominates the scene.

C.L. Mowat (via Dave Waldren)

Rhagair - Foreword

The harbour at Porthmadog was the lower terminus of several slate-carrying railways, of which the Festiniog is rightly the most famous, and the most thoroughly researched. The lesser systems, however, also deserve proper recognition, and the Croesor now has found its chronicler in David Southern, ably assisted by Adrian Barrell. Built during the boom years of the 1860s, it made use of level stretches of track interspersed with some spectacular inclined planes to reach remote quarries on the slopes of Cnicht and Moelwyn Mawr. Its builder was Hugh Beaver Roberts, a Holyhead captain's son who became a solicitor, an election agent for the Penrhyn family and the Conservative interest, and rich enough to send his son to Eton. It probably repaid his investment; it was still carrying slate into the 1930s and there was some traffic in agricultural manure into the late 1950s. It enabled a remote upland cwm to become, in its own way, a remarkable and innovative industrial culture.

Adnabyddir Cwm Croesor fel ardal chwarelyddol, fel mae'r llyfr hwn yn dangos, ond oedd cyfoeth amaethyddol y cwm a'r cylch yn cynnal bechgyn y fro hefyd. Bu'r bardd a llenor Glaslyn yn was bach yn Ynysfor cyn iddo fynd i Ffestiniog i fod yn chwarelwr. Bu Bob Owen yn gweini ar ffermydd yn yr ardal am dair blynedd cyn cael swydd fel clerc yn chwarel Parc a Chroesor. Yn gasglwr llyfrau ac arbenigwr ar achyddiaeth Gymreig, dywedir ei fod wedi treulio ei fis mêl yn y Llyfrgell Genedlaethol. Mae ei lyfr Diwydiannau Coll yn adlewyrchu'r hen ddrefn wledig ond mae'n cynnwys hefyd y grefft o griwlio inclêns. Un arall oedd Moses Kellow, rheolwr a pheirianydd y chwarel, a gyflwynodd nifer o ddatblygiadau pwysig yn y diwydiant llechi yng Nghymru, i dyllu'r graig ar gyfer ffrwydro, ac i symud clytiau a gwastraff o fewn y gwaith.

The story of this fascinating railway and the community which it not only served but created, is told here.

David Gwyn
2018

Preface

The Croesor Tramway follows the approach I adopted for *The Bryngwyn Branch*, published in 2014 by the WHRHG [1]. It looks not only at the history of the tramway, why it was built and some of the people involved, but is also an exploration of the physical remains of the tramway and the quarries and community it served.

Like *The Bryngwyn Branch*, co-authored with the late John Keylock and edited by David Allan, this book is the product of much collaboration with others as will be evident from the Acknowledgments.

Over a number of years, I have explored the line and the nearby quarries all the way from the Rhosydd Quarry to the quays on the harbour at Porthmadog, before the Ffestiniog & Welsh Highland Railways re-opened the section between Croesor Junction and Porthmadog.

The reinstated railway is more than just a seasonal tourist attraction. It provides not only a car free means of appreciating Snowdonia but also takes the walker to a variety of destinations including the foot of Snowdon. Such a facility was first provided by the North Wales Narrow Gauge Railways Company as long ago as 1881 when its line from Dinas (on the LNWR's standard gauge line from Carnarvon to Afon Wen) to Rhyd Ddu was completed. The story of the Croesor Tramway is closely linked with that of the NWNGR and its subsequent reincarnation as the Welsh Highland Railway in 1922 and, much more recently, the 'new' WHR.

My hope is that this book will be of interest not only to the 'enthusiast' but also to the general reader, shedding light on a relatively obscure tramway that served the quarries and community of Cwm Croesor.

Dave Southern
Conwy, North Wales
2018

[1] *The Bryngwyn Branch* D. Southern with J. Keylock, WHRHG, 2014 - ISBN 978-0-9930821-0-8

Acknowledgements

In preparing this book, I have had recourse to many different sources of research and these are listed in the Bibliography and referenced in the footnotes. I thank the authors of these books and articles for their research into the intricate story of the Croesor Tramway and the associated railways, tramways and quarries. The establishment of the Welsh Highland Railway Heritage Group in 1997 stimulated much new research and exploration and the Group's Journal *Welsh Highland Heritage (WHH)* has been a rich source of new information in the excellent articles by a number of different authors.

Adrian Barrell, the researcher into 'all things Croesor', made available the extensive archive accumulated over many years by him and others from delving into the history of the Cwm Croesor Quarries. Without that, this book would not have been possible in its present form and in particular the chapter of reminiscences of those who knew the tramway when it was working. Adrian has also made constructive comments on the initial drafts, for which I thank him.

I also thank my publishers, the Welsh Highland Railway Heritage Group, who took on the task of preparing my manuscript for publication, including the chairman, Nick Booker, and Peter Liddell, the Journal editor, who have been exemplary editors in suggesting additions and changes to the structure and contents. Peter has done his usual splendid job in selecting photographs, writing captions and designing and preparing the book for publication.

Photographs, illustrations and maps are drawn from a variety of sources including the WHRHG's archive with the help of Dick Lystor and Peter Liddell. David Allan has also provided a rich source of material. The maps and plans drawn by the late J (Jim) M Lloyd that originally appeared in J I C Boyd's *Narrow Gauge Railways of South Caernarvonshire* are reproduced with the kind permission of the Welsh Railways Research Circle and the current owner of The Oakwood Press, Stenlake Publishing. I thank both Dave Sallery and Iain Robinson who, like me, have spent many happy and wet hours walking and photographing the remains of a once great Welsh industry.

Many of the 'modern' photographs are from my own camera, taken over the many years that I have explored the Croesor Tramway and its quarry customers.

Where known the appropriate acknowledgements, copyrights and attributions are given. In some cases, it has not been possible to identify copyright ownership, but any information regarding such ownership would be welcomed.

Richard Maund, David Gwyn and Michael Bishop have been critical friends and made many valuable suggestions for improving the various iterations of the manuscript. I thank them for their contributions, patience and knowledge.

However, as the author, I must ultimately take responsibility for any inaccuracies, errors or omissions.

Dave Southern

Conwy, North Wales

September 2018

Miscellanea

Abbreviations

AWCR	Aberystwith and Welsh Coast Railway
BoT	Board of Trade (later Ministry of Transport)
FR	Festiniog Railway (still the railway's statutory title)
LRO	Light Railway Order
MANWEB	Merseyside and North Wales Electricity Board - the nationalised regional electricity supplier and distributor for Merseyside, North Wales and parts of Cheshire; now part of Scottish Power
NWNGR	North Wales Narrow Gauge Railways
PB&SSR	Portmadoc, Beddgelert and South Snowdon Railway
PC&BTRC	Portmadoc Croesor and Beddgelert Tram Railway Company
Power Co.	North Wales Power and Traction Company Limited (sometimes referred to as NWPT in documentation)
WHR	Welsh Highland Railway

Spelling

Spellings are appropriate to the historical context, i.e. 'anglicised' spellings pre-1973, in particular where they appear in an historical context such as a document or in the title of a company. Welsh spellings are used in any post-1973 context. There are exceptions, e.g. Carnarvon/Caernarvon/Caernarfon, where there were different 'anglicised' versions of the name. Caernarfon spelling: prior to 1926, the spelling of town and county was Carnarvon. From 2nd February that year the town became Caernarvon, the L.M.S. station following suit on 27th March and the county from 1st July. The current spelling was not adopted until well after the Second World War.

Measures

Measurements / Dimensions are provided as per the original source, usually in Imperial form.

100 yards = 91.44 metres

1 mile = 1.61 kilometres

22 yards = 1 chain

A chain measures 66 feet, or 22 yards, or 100 links, or 4 rods. There are 10 chains in a furlong, and 80 chains in one statute mile. An acre is the area of 10 square chains (that is, an area of one chain by one furlong).

To express the acre in metric terms, 1 acre = 0.404685642 Hectare

Where 'chainages' are quoted in this work they are measured from the tramway's 'zero point' which lay adjacent to the back of the Oakeley Wharf at Portmadoc Harbour. This was located some distance from the Festiniog Railway 'zero point'.

Additionally, when measuring above the various inclines, distances are measured in the horizontal plane, i.e. as would be viewed on a map, and do not take account of the additional track length due to the angle of the inclines. This 'error' amounts to less than 200 feet between Harbour and the Croesor Quarry winding house.

Access

Please bear in mind that all tramways and quarries are private property and some may still be working sites. Appropriate permission should be sought before a visit. Common sense is necessary here, a stroll along the unfenced parts of the Croesor Tramway or around the surface of Rhosydd Quarry, hardly needs permission, but a visit to the Parc Mills area does.

The usual mountain walking procedures and precautions apply. It is dangerous to venture underground, on to waste tips or to attempt to walk up or down the inclines. Underground exploration should only be undertaken by those suitably equipped and experienced and, most importantly, never alone.

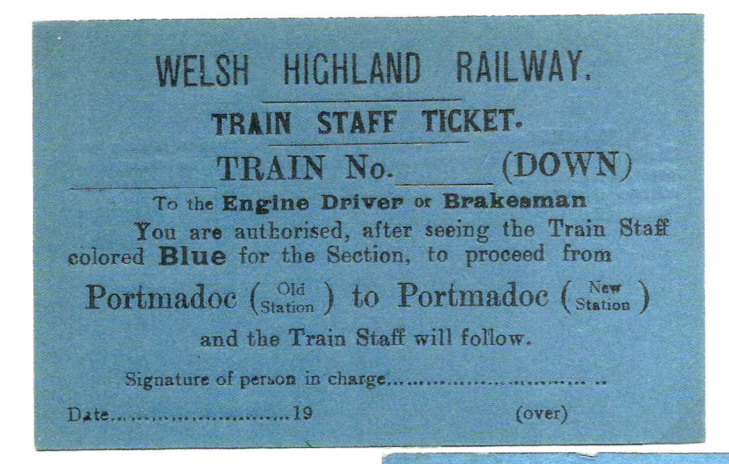

Sample WHR Train Staff ticket covering the section from Portmadoc Old to Portmadoc New

Sample WHR Train Staff ticket covering the section from Portmadoc New to Croesor Junction

Sample WHR Train Staff ticket covering the section from Croesor Junction to Beddgelert

Welsh Highland Railway Train Staff Tickets - the operational arrangements in place covering the operation of Croesor traffic after the opening of the Welsh Highland are discussed in Chapter 4. As explained there, the latter two of these tickets represent the period after the beginning of June 1924.

Chapter 1

Background

In the nineteenth century the slate quarries of Gwynedd in North Wales not only constituted the most important industry of the region, making a dramatic contribution to both the economy of North West Wales and its landscape, but they also supplied the bulk of the world's roofing slate and architectural slate slabs. The quarries grew on a gigantic scale in many cases; huge quantities of stone were removed from terraced hillsides, pits and underground mines and because of the very high wastage rate (90%−), the rubble tips soon came to dominate the slate producing areas, forming the most potent visual reminder of the industry. The industry's expansion from the late eighteenth century onwards contributed to a regional population explosion (the population of Carnarvonshire increased from 41,500 in 1801 to 126,900 in 1901) which in turn resulted in changes to the landscape peripherally rather than directly related to mineral extraction. For example, profits from slate extraction enabled the Douglas-Pennant family to build the neo-Norman Penrhyn Castle [1], while the higher wages offered by the quarries attracted workmen and their families off the land, to make their homes in the new villages and on the small holdings which sprang up on remote mountainsides [2].

High up in remote Cwm Croesor (Croesor Valley) were the Rhosydd, Croesor, Pant-mawr and Fron-boeth Quarries, while further down the valley lay Park Quarry, located near the Parc inclines which carried the tramway from the Croesor Valley floor down to the Glaslyn flood plain near Gareg-hylldrem.

Although the term *quarry* is generally used to describe such slate workings, strictly a true quarry comprises only surface workings. Many of the slate workings in this area were underground and therefore more correctly should be described as *mines*. Nevertheless, in line with common usage, we will use the general descriptor *quarry* including, of course, in those instances where the word was part of the undertaking's legal title.

The Park and Croesor Slate Quarries, which came under common ownership in 1895, were upwards of 2,000 acres in extent and described by *The Slate Trade Gazette* of February 1911[3] as having 'the finest range of slate-bearing strata in the world' with a length of 312 miles.

The Ffestiniog series of veins, comprising in ascending order *The New* or *Deep Vein*, *The Old Vein*, *The Small Vein*, *The Back Vein*, and *The North Vein*, and lying immediately on a thick stratum of felspathic porphyry, were known locally as *The Glanypwll Trap*. They traverse both properties for their entire length and in 1911 represented a slate reserve which, at a conservative estimate, amounted to over two thousand million tons, of which over six hundred millions consisted of the world-famous *Old Vein* slate [4]. At Croesor Quarry, the *Old Vein* was accessible, dips down at 27 degrees, and is around 120 feet wide at right angles to the bedding plane and 300 feet horizontally.

---

1 Penrhyn Castle was created between about 1822 and 1837 to designs by Thomas Hopper, who expanded and transformed an existing building beyond recognition. Hopper's client was George Hay Dawkins-Pennant (1763-1840), who had inherited the Penrhyn estate on the death of his second cousin, Richard Pennant, who had made his fortune from slavery in Jamaica and from local slate quarries. The eldest of George's two daughters, Juliana, married Grenadier Guard, Edward Gordon Douglas, who, on inheriting the estate on George's death, adopted the surname of Douglas-Pennant. www.douglashistory.co.uk/history/families/douglas-pennant

2 A. Davidson, A. Jones, G. P. Jones & D. Gwyn, 1994

3 The *Slate Trade Gazette* of February 1911 / http://www.llechicymru.info/croesor.english.htm (accessed 10 October 2018)

4 ibid

Background | The Croesor Tramway

The Croesor Quarry was first worked in the 1840s and despite extensive investment failed to produce any significant volumes of slate and closed sometime in the late 1870s/early 1880s. However, in 1895, the quarry reopened under the direction of Moses Kellow, 'a fearless innovator' [5] who set about modernising working practices and methods. By 1911, it was being worked on an extensive scale exclusively for *Old Vein* roofing slates.

The Parc Quarry was opened in about 1851 and was chiefly worked for slate for manufacturing purposes, such as slate ridging, slabs, etc., the perfect straightness of the cleavage, which characterised the rock of the quarry, making it especially suitable for this kind of product.

Figure 1.1

The Croesor Slate Quarry framed by the mass of Moelwyn Mawr

Dave Sallery 2006

The two quarries complemented one another, enabling The Park and Croesor Slate Quarries Co. Ltd to supply its customers' needs in roofing slates, slate ridging, slabs, or any other variety of slate product. Peak output under Kellow's direction was between 5,000 and 6,000 tons per year, though it declined in the later years, until the quarry closed in 1930 [6]. From the late 1940s until the early 1970s, Cooke's Explosives used the chambers to store propellants but this ceased when the Central Electricity Generating Board feared that an underground explosion would damage the dams of the nearby Ffestiniog pumped storage power station. The quarry was worked on seven levels and there was an underground link to the nearby Rhosydd Quarry.

Rhosydd Quarry is four miles from Blaenau Ffestiniog and eight from Porthmadog. The main surface workings are at Bwlch Cwm-orthin, a high pass separating the valleys of Cwm Croesor and Cwm Orthin. Small scale working of the site began in the 1830s, but was hampered by the remote location and the lack of a transport system to carry the slates to market. Transport was made more difficult by the attitude of the owners of Cwm-orthin Slate Quarry, through whose land the most obvious route to the Festiniog

5 Lindsay, Jean 1974
6 ibid.

Railway ran. The solution was found in 1864, with the opening of the Croesor Tramway, to which the quarry was connected by a long single-pitch incline. The quarry passed through various financial crises and prospered for a while, but profitability declined, and in 1900, a large section of the underground workings collapsed. Following the onset of the First World War, the quarry was mothballed but reopened in 1919. It was then bought by members of the Colman family, better known for mustard. They kept it running until 1930, but failed to find markets for the finished product. In 1948, the site was abandoned.

Figure 1.2

The final approaches to Rhosydd Quarry along the ½ mile trackbed of the tramway leading from the head of the Rhosydd incline. The rubbish tips are clearly visible on the skyline, often the only indication of 'quarrying' activity where the workings were predominantly underground.

Dave Sallery 2007

In the early years of working, getting the output of the quarries to market was problematical. In the case of the Croesor Quarry, the slate was carried by pack mule over to the adjacent Cwm Orthin and down to the Festiniog Railway at Tan-y-grisiau. However, anecdote [7] has it that an early pre-tramway exit route for Croesor slate was down Cwm Croesor to Llanfrothen and then direct by road to Penrhyn. The long straight road from Llanfrothen to Penrhyn was said, to have been built for that purpose by the Festiniog Railway.

The Croesor Tramway was conceived as a direct connection from the quarries to the sea at Portmadoc. Under the guidance of Hugh Beaver Roberts and two other quarry proprietors, work on the Tramway was started in 1862. It opened to goods and mineral traffic in 1864. The Rhosydd Quarry at the head of the valley was connected that same year.

The tramway continued to carry slate from the quarries along Cwm Croesor until 1944, when the last wagons were sent down the Rhosydd incline and on towards Portmadoc. The track from Croesor Junction to Croesor Village was lifted between 1944 and 1953. The remaining track, between Croesor and Blaen-y-cwm continued in unofficial use by local farmers until the late 1950s [8].

---

7 Adrian Barrell, August 2018 - in note to Nick Booker
8 Boyd, Vol 1 (1988) p. 123

Historical Summary

Croesor Tramway completed and opened to goods and mineral traffic through to Beddgelert Siding weigh-house	1864
Rhosydd Quarry connected to Croesor Tramway	1865
Croesor & Portmadoc Railway Company incorporated, enabling extension to Portmadoc Harbour	1865
Croesor United Quarry Co Ltd formed	1866
Rhosydd Company fails – New Rhosydd Slate Company Ltd formed	1873 - 1874
Croesor United Quarry Co – into liquidation	1874
Croesor New Slate Company Ltd formed	1875
Portmadoc, Croesor & Beddgelert Tram Railway Company (promoted by C&PR) incorporated - PC&BTR acquired the C&PR	1879
Croesor New Slate Co ceased operation	1882
PC&BTR into receivership	1882
Croesor Quarry became part of Park & Croesor Slate Quarries Ltd - Worked by S Pope	1895
PC&BTR Co sold to Portmadoc Beddgelert & South Snowdon Railway Company Ltd	1901
New Rhosydd Slate Company Ltd bought out and worked until 1930	1921
PB&SSR, and hence lower portions of the tramway, incorporated into the Welsh Highland Railway	1922
Croesor and Rhosydd Quarries - the last to be closed	1930
Welsh Highland Railway closes	1937
WHR partly dismantled	1941
Croesor Tramway between Junction and Village dismantled	1944 - 1953
Railway re-instated along tramway route between Croesor Junction and Porthmadog as part of resurrection of the WHR	2010 - 2011

Chapter 2

History of the Croesor Tramway

Birth and development

The Croesor Tramway was built to the order of Hugh Beaver Roberts (1820-1903), a solicitor from Bangor. No parliamentary powers were needed to build the tramway, as Roberts owned much of the land. On the sections he did not own, way leaves [1] were obtained to secure the right of way. Roberts had interests in a number of North Wales quarries and in the Croesor valley. He owned the land on which several quarries were located, including the Croesor Fawr Slate Quarrying Co Ltd of which he was a director.

Charles Easton Spooner surveyed the route in 1863 and at Portmadoc agreement was reached to use the Festiniog Railway tracks around the harbour.

The exact date that work started on the tramway is not recorded, but on 6th December 1862, the *North Wales Chronicle* reported:

> *"The Croesor Slate Quarry, which is situate on the south-west side of the celebrated Moelwyn Mountain, has been pretty fairly developed; but the extreme difficulty of removing the slates to the Port for exportation has hitherto operated as a very serious drawback. In order to obviate this difficulty, it was resolved sometime ago to make a tramway to Portmadoc harbour, past Ynysfawr and along the flat ground of Traeth Mawr, so as to join the inner harbour by Ynystowyn; and Mr. Spooner, the civil engineer, was directed to survey the line, and to supply the required plans and sections. This having been done to the satisfaction of the Company, it was resolved to carry out the design; and we are glad to say that last week the undertaking was fairly commenced, at the end next the quarry"*

The Croesor Tramway was constructed by Pritchard and Gregory of Bangor. The tramway was of 2 ft gauge on wooden sleepers using low weight bulb tee section rail in cast iron chairs with a wider chair at the joints as there were no fish plates. Rails were 20 lb per yard in 15-foot lengths suitable for horse traffic and vertical-boilered engines.

Figure 2.1

Croesor Tramway rail recovered from near Croesor Village in 1961.

The views show the T-bulb rail section and two types of chair - one plain and the other extra-wide to support rail joints

Edward Dorricott

By April 1863 good progress was being made with the construction of the tramway trackbed and a large delivery of rails arrived ready for track laying [2]. A letter from

---

1 Way leaves - A right of way granted by a landowner, generally in exchange for payment and typically for purposes such as the erection of telegraph wires or laying of pipes. 'companies must have way leaves for work they want to carry out on private land' 'they must seek a way leave' – Oxford Dictionaries

2 "The Trade of the Port". *The North Wales Chronicle and Advertiser for the Principality.* 4 April 1863.

Benjamin Wyatt in the *North Wales Chronicle* of 2nd January 1864 states that during the last ten months of 1863 some 600 tons of slate had been conveyed by road from Croesor Quarry to Penybwlch [*sic*] siding (Penrhyndeudraeth) for transhipment there to the Festiniog Railway for onward transport to the port.

It is likely that upper sections of the tramway were brought into use as they became available to minimise the necessary road element of the journey – Boyd (Vol. 1), p. 100, states that such use began late February 1864. The *North Wales Chronicle* of 16th July 1864 reported that the line was completed – by "the large bridge near to the Glaslyn Inn" (Pont Croesor) - on 7th July 1864, so operation over the whole line could then have started. An "official" opening date of 1st August 1864 was quoted in *The Railway Yearbook* (e.g. for 1910) for the PB&SSR - a detail which presumably originated from the railway itself. Boyd (Vol. 1), p. 117, states that the first wages paid (presumably – although he does not specify – for operating the whole tramway) were for work performed from that 1st August date.

Roberts incorporated the Croesor & Portmadoc Railway Company (C&P) by an Act of Parliament of 5th July 1865, with a registered office in Carnarvon. The company had share capital £25,000 in £50 shares.

The Act was *"to provide for the maintenance and use by the public of the existing railway made by Hugh Beaver Roberts of Plas Llanddoget which commences near the rock or place known as Carrig Hylldrem in a certain field called Cae Ochor Rhainwal, part of the farm called Park and terminates at or near Ynys Cerrigduon at Portmadoc in the parish of Ynyscynhaiarn, together with station sidings and works and to adapt and use for passengers as well as other traffic."*

Under the Act's powers, the tramway's gauge could be increased from 2 feet to any gauge not exceeding 4 feet 8½ inches.

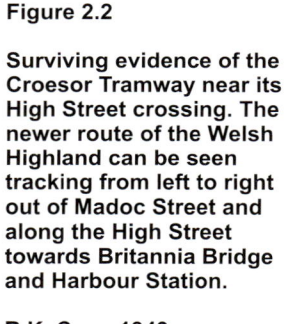

Figure 2.2

Surviving evidence of the Croesor Tramway near its High Street crossing. The newer route of the Welsh Highland can be seen tracking from left to right out of Madoc Street and along the High Street towards Britannia Bridge and Harbour Station.

R.K. Cope 1949
via Roger Carpenter

Not only did the Act provide statutory powers both for the lower part of his existing tramway and for a 1 mile 19 chain extension to Portmadoc wharves and to Borth y Gest, the latter part of the extension never being built, but there were also powers to build a branch from Ynysfawr towards Beddgelert to serve the various mines that had opened around the village. This proposed branch was not built, although the Welsh Highland Railway did eventually construct the route from Ynysfawr to Beddgelert.

The 2½ mile section north east of Gareg-hylldrem, owned by Roberts and within the Croesor valley including the two inclines, was not covered by the Act.

The C&P Plans & Sections were drawn by Charles Easton Spooner, described as Manager of the Festiniog Railway Co. These show the terminus of the existing tramway as being just the town side of a weighing machine alongside where Beddgelert Siding was later located, about midway between the standard gauge crossing and Pen-y-mount. This spot also marked the commencement of the extension into Portmadoc and the wharves.

Maps dating to the early days of the Tramway mark this location as *Beddgelert Siding*. However, while that name was used in the contract dated 30 April 1922 with McAlpines, from the start of Welsh Highland services, the name changed and the later name *Gelert Siding* is referenced in the *Official Hand-book of Stations* [3]. *Gelert* is the name used by S E Tyrwhitt, the WHR's General Manager, in his communication with the MoT in May 1923 and in a memo in September 1925. *Gelert's Sidings* is shown in the WHR's 1924 working timetable and *Gelert Interchange Sidings* on a plan supplied for the 1926 MoT inspection. Therefore reference to either *Cambrian* or *Beddgelert Siding* in the Welsh Highland era would be incorrect.

It is perhaps worthy of note that, despite these factors, the quarries generally referred to these facilities as *Croesor Sidings*!

The C&P promoted an Act in 1879 to build an extension north to Beddgelert under the name of the Portmadoc, Croesor & Beddgelert Tram Railway Company (PC&BTR) with a capital of £23,000. However, the project failed to attract sufficient money or public interest. Thus in 1882, James Cholmeley Russell, a London based barrister, financier, property developer and railway entrepreneur and the Receiver and Chairman of the NWNGR became a mortgagee of the PC&BTR. The role gave him effective control and, in the same year, the company was put into receivership by another mortgagee, thereby blocking any plans for the transformation of the tramway for many years to come.

In 1901, its assets and powers were sold to the Portmadoc Beddgelert & South Snowdon Railway Company for £10,000 with the purchase confirmed by the PB&SS Railway Company Act of 17th August 1901.

The PB&SSR's purchase was to enable it to maintain the section of the tramway between Portmadoc and a proposed junction with a new section of line to Beddgelert and a continuation thence up Nant Gwynant as far as the outlet from Llyn Gwynant.

The Company's grand plan included proposals to operate its railways by electricity, working closely with the North Wales Narrow Gauge Railways and Festiniog Railway

---

3 *The Official Hand-Book of Stations* was a large book listing all the passenger and goods stations, as well as private sidings, on the railways of Great Britain and Ireland. Last published in 1956 by the British Transport Commission (under the Railway Clearing House name) it provides an historical snapshot of the railways of the time.

companies and to construct a new hydro-electric power station in Nant Gwynant. The full story is told in *Ghosts of Aberglaslyn*.[4]

The 1872 Parliamentary Act that set up the North Wales Narrow Gauge Railways included proposed running rights over the Croesor & Portmadoc Railway. The line proposed by the NWNGR would have formed a junction with the Croesor Tramway close to Gareg-hylldrem - near or at the point where the WHR would eventually meet the Croesor in 1922.

On completion of the extension to Portmadoc, the Croesor was connected to the Festiniog Railway at Portmadoc harbour, where it joined the fan of lines to the slate wharves. From the earliest days, it is probable that Festiniog Railway wagons traversed the Croesor. Certainly by 1873 the practice had been formalised, with the Festiniog charging the quarries along Cwm Croesor a tonnage fee for use of their wagons on the tramway.

The track was owned by the tramway with the wagons being owned by the Quarry owners - the horses were supplied by contractors.

Figure 2.3

Parc and Croesor wagon number 196, now part of the Welsh Highland Heritage Railway wagon collection.

Dave Southern

The wagons were built in an unusual way in that the axle box castings were bolted directly to the angle iron framing on the bottom of the body, as there was no under-frame. Some wagons were braked with a handle on the side attached to blocks on both pairs of wheels. This type of wagon could be used on the return to Croesor and the quarries for coal, bricks and quarry stores. Farmers used the tramway to bring supplies to their farms where there was no road access.

The final stages of the route through Portmadoc to the Harbour were latterly shared with the Gorseddau Tramway. This was originally a 3 ft narrow gauge railway built in 1856 to link the slate quarries around Gorseddau with the wharves at the harbour at Portmadoc. It was a forerunner of the Gorseddau Junction & Portmadoc Railways Co. and subsequently, by virtue of its route through Portmadoc, the Welsh Highland Railway. The first section was a horse worked tramway that ran for 1½ miles from Portmadoc Harbour to Tremadoc Ironstone mines at Llidiart Yspytty (between Tremadoc and Penmorfa) and was in use by 1845. The 3 ft gauge line was extended for 6½ miles from Tremadoc to Gorseddau Quarry (in Cwm Ystradllyn) and in this longer form was in use by May 1857 and known as the Gorseddau Tramway. By Act of Parliament the section of the Gorseddau Tramway between Portmadoc Harbour and Braich-y-bib (also known as Braich-y-big - the name of the nearby farm) was converted to 2 ft gauge and extended

4 Manners and Bishop, 2016

from there to Cwm-trwsgl and the whole statutory undertaking formed The Gorseddau Junction & Portmadoc Railways. The 'new' railway opened to public goods and mineral traffic in June 1875. There is no evidence to suggest that the original line to the Gorseddau Quarry was converted to the 2 ft gauge as the quarry had closed in 1867. The Croesor Tramway originally crossed the Gorseddau on the level as the latter exited Madoc Street, the line of its original route, before crossing Portmadoc High Street. However, on re-gauging, the Gorseddau was re-routed along the bank of y Cyt to make junction with the Croesor close to the Flour Mill. Both railways served the same wharves.

The Gorseddau tramway was disused by 1892. The most prominent existing feature on the line is the roofless slate mill at Ynys-y-pandy. The line followed y Cyt from Portmadoc to Tremadoc. This had been created as a land drain as part of one of Madocks' early reclamation schemes. It has been said that it was once used as a canal for a 100 ton ship.[5] It is crossed by the Cambrian Coast railway line adjacent to the car park of the present Welsh Highland Heritage Railway.

Figure 2.4

Rails stacked at the Portmadoc end of the Gelert Siding complex, collected as the southern section of Welsh Highland was lifted.

The last surviving sections of the Croesor were to be lifted four years after this photograph was taken.

R.K. Cope 1945
via Roger Carpenter

Declining years and closure

In 1920 Parc Quarry closed, followed by Fron-boeth, Pant-mawr, Croesor and Rhosydd in 1930. With the closure of the quarries, regular traffic ceased but small amounts of traffic continued and the upper section was used on a casual basis by local farmers until the 1950s.

In 1926 the Portmadoc High Street section of the tramway was lifted as its Croesor rail was unsuitable for WHR locomotives and any freight was taken to the wharves via Harbour station, rather than direct. The Croesor section of the Welsh Highland Railway remained after the rest was lifted in 1942 to allow for the possible resumption of quarrying in the Croesor valley and was not finally removed until 1948/49. Up until then, the tramway remained in working order with a gang from Croesor quarry visiting the line a few times a year to maintain it. In 1943, the Blaencwm pipeline needed repair, so the army used

5 Festipedia, at https://www.festipedia.org.uk/wiki/Y_Cyt (Accessed 10/10/2018)

the track between Croesor village and Blaencwm to aid and support this work. The pipes came from the Midlands by rail to Penrhyndeudraeth station on the Cambrian Coast line then by road to Croesor where they were loaded onto rail wagons to be taken to Blaen-y-cwm Incline. This traffic marked the only occasion that trains on the upper section used a diesel locomotive [6], all other traffic was horse drawn. In addition, there was still fuel oil traffic from Croesor village three times a month to Rhosydd quarry via the incline.

The track was removed from the inclines in the Summer of 1950 by W O Williams of Harlech. The valley section was finally lifted around 1970. The section from Croesor junction to Gelert Siding was lifted between August 1948 and August 1949 and rails were stacked at the Siding.

Figure 2.5

The contractor's Simplex left to nature near Portmadoc New after track lifting was complete.

Other photos show a right hand end panel sign written *'Shanks & McEwan Contractors'*. It was presumably acquired from them by Williams of Harlech prior to the lifting of the Croesor Tramway. The FR subsequently cannibalised parts for use on *Mary Ann*

D. Rendell 1952

The motive power used by the Contractor was a Motor Rail petrol tractor, No 14, which was subsequently abandoned at Portmadoc New station north of the standard gauge and it was cut up around 1955. Due to the contractors' lack of access to the track south of the standard gauge following the lifting of the crossing over the GWR line in December 1937 [7], the Cross Town Link section was not finally removed until May 1949.

There are no formal closure dates for the Tramway, with the exception of the section operated by the WHR, which closed to passengers after the last trains ran on 26th September 1936 and to freight from 31st May 1937 (this latter being the official closure date).

The final workings on the Croesor Tramway section under WHR ownership

After regular traffic ceased, with closure of the Welsh Highland Railway, occasional trains did run along the section from Croesor Junction to Portmadoc, but only as far as Portmadoc New north platform after 1937, before the line was lifted.

6 Hawkes, date unknown
7 *WHH* 58 p. 8

On 19th June 1937, the WHR Hunslet 2-6-2T *Russell* left Dinas for Portmadoc to collect any FR owned wagons and to return them to their parent railway. On return *Russell* collected the Welsh Highland Railway's ex-WDLR [8] Baldwin *No. 590* from Boston Lodge together with any WHR wagons remaining between Harbour station and Dinas.

On 14th August 1941, a tractor with a bogie flat and slate wagon ran through to Portmadoc New north platform to test the line ahead of the planned demolition trains. Demolition of the Welsh Highland reached Croesor Junction from Beddgelert in June 1942.

Figure 2.6

Three Croesor wagons being recovered by the F.R. The wagons are seen near Bryn-hyfryd being moved towards Croesor Village.

A.G.W. Garraway

The final traffic to use the lower section of the tramway occurred when the upper parts of the Welsh Highland line were requisitioned for scrap and the recovered materials were removed via the Gelert Siding. This process was repeated when the Croesor junction to Portmadoc section was lifted from August 1948 to August 1949 by W O Williams of

Figure 2.7

The same location somewhat earlier, judging by the vegetation growth, with a couple of abandoned wagons. Note the flatbed wagon with the de-mountable wooden body.

D. Mitchell

---

8 War Department Light Railways - the British Army's military railway command in World War 1.

Harlech. The section from Croesor Junction had been left in situ when the WHR was lifted pending the possible reopening of the Croesor quarries, which never happened.

After the removal of scrap from the Croesor tramway, the standard gauge line to Gelert Siding was closed and the rails were finally lifted ca. 1959.

As for rolling stock, remains of three solid sided wagons were reclaimed from the tramway above the village by Festiniog revivalists and used for carrying ash at Boston Lodge; these have now moved to the Bala Lake Railway. The sidings, which Croesor traffic had used, remained at Portmadoc harbour and saw the storage of FR wagons until 1949.

Figure 2.8

A solitary cow gazes on the desolate scene as the tramway quietly reverts to nature.

The image shows the trackbed by the access to Croesor-fawr farm, looking towards Blaencwm

F.R. Archives

Figure 2.9

More cows enjoy the evening sun along the tramway trackbed

F.R. Archives

Chapter 3
The Route Described

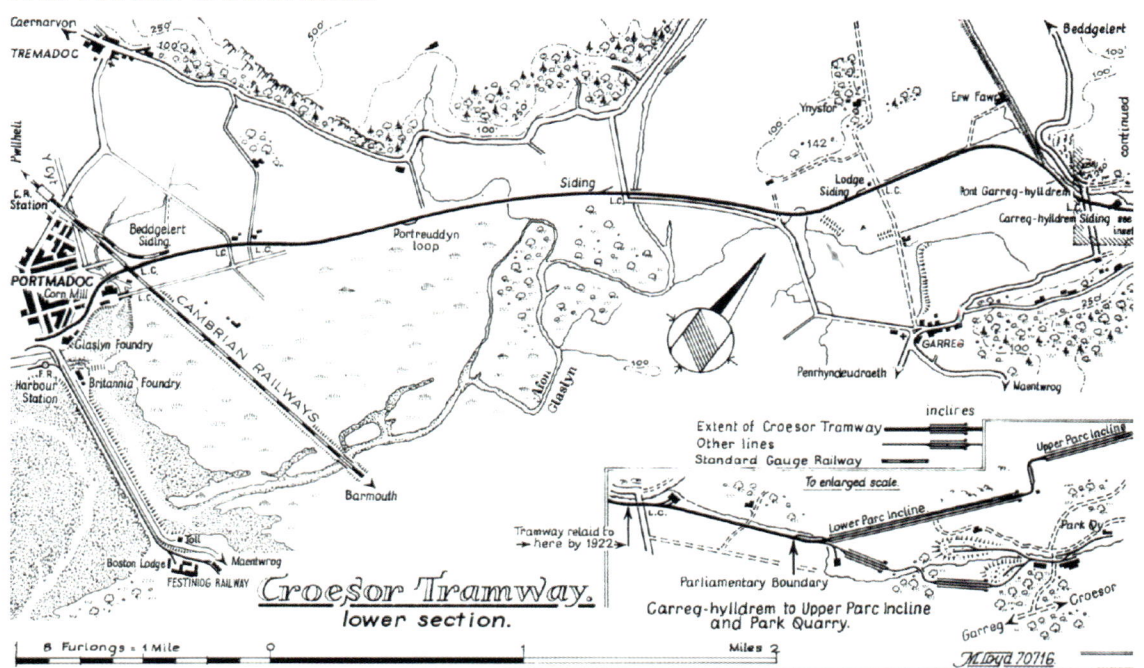

Figure 3.1
The route of the Croesor Tramway from Portmadoc to Carreg Hylldrem.
J.M. Lloyd

Portmadoc to Gareg-hylldrem

The tramway ran from the quays at Portmadoc Harbour, commencing at the rear of the Oakeley Slate Company's wharf. It was linked with the Festiniog Railway's line near a timber yard at Cornhill and was connected to the public wharf, Oakeley, Greaves and Rhiwbryfdir Wharves.

Figure 3.2
Portmadoc Harbour looking towards the sea. The open area beyond Britannia Bridge was the Welsh Slate Co. Wharf and the long sheds beyond mark the Oakeley Wharf. The Croesor Tramway entered on the right and passed in front of the two sheds to make a physical connection with the FR behind the Oakeley's facilities.
iBase 1895

On leaving the harbour area, the line swung to the north and ran through the western edge of the town, crossing Portmadoc High Street at right angles on the level. The Gorseddau Tramway of 3 feet gauge ran to its right-hand side and crossed the Croesor Tramway by a rail level crossing. There were dual track gauges in the harbour area until the Gorseddau was re-gauged (Chapter 2). After crossing the High Street, the Croesor passed behind terraced houses on Madoc Street before curving around the edge of the inner harbour and then eastwards across Snowdon Street towards the Flour Mill (later Snowdon Mill) and the junction with the Gorseddau Tramway, installed when the latter was re-gauged. Here a fixed roof provided cover for the mill's loading operations.

Figure 3.3

Leaving the Harbour complex, the Tramway crossed the High Street and turned into Madoc Street en-route to the flour mill and the Cambrian Crossing.

The fan of tracks between the Oakeley Wharf and the Wood Yard are clearly visible here, with the Croesor line to the left.

Commercial Postcard via Martin Pritchard

Figure 3.4

The Portmadoc Flour Mill viewed from the area of the North Wales Slate Co. Works, looking across the pair of Gorseddau tracks in the foreground. The nearer track was a siding for the slate works with the Gorseddau 'main line' beyond.

Note the awning covering the loading area for Croesor traffic and the horse at the head of a train loaded with flour sacks in preparation for departure towards Croesor.

WHRHG collection

Figure 3.5

Slate stacks alongside the Croesor Tramway at Beddgelert Sidings. Note the standard-gauge wagons beyond the slate stacks and the Welsh Slate Company's chimney in the distance.

Park and Croesor Quarries Handbook.

Beyond the mill the tramway passed the future site of Portmadoc New station to reach a flat crossing on the Aberystwith & Welsh Coast Railway. The A&WCR was absorbed into the Cambrian Railways in 1866. This level crossing, with its two separate gauges of track, was controlled by a signal cabin and was fitted with trap points on the narrow gauge either side of the crossing (at 0 m 39 ch).

Beyond the A&WCR, the line passed Beddgelert Siding, where there was an interchange with a short standard gauge branch from the A&WCR at Portmadoc Station. Much of the slate from the quarries was transferred here to the Cambrian for shipment onwards. There was a slate stock yard, weighbridge and sidings. Further on, at 66 chains, was a siding for loading sand on the east side. From here the line ran north east across Traeth Mawr along Creassy's Embankment, built by William Madocks in 1800 to cut off the north-west of Traeth Mawr, the estuary of River Glaslyn, which allowed him to establish the town of Tremadoc. The Embankment is named after James Creassy, the waterways engineer and surveyor employed by Madocks to supervise its construction, which was primarily of earth and sand.

Figure 3.6

Slate being unloaded from the Croesor Tramway at Beddgelert Sidings.

Here was the Tramway southern terminus before the link to Portmadoc Harbour was completed.

Standard gauge wagons can be seen beyond the slate stacks.

Slate Trade Gazette Feb 1911

The tramway ran to Portreuddyn (1 m 48 ch) where there was a passing place, not retained when the WHR was established. The tramway continued on a low embankment to Pont Croesor. The tramway crossed the Afon Glaslyn by an eight-span bridge on stone pillars with, latterly, a road alongside. Crossing the Glaslyn had been problematic prior to the building of this bridge, the options being a long detour or use of ferry services at points along the river. After the bridge was built it was finally possible to build a road to Llanfrothen from the old coast road near Pren-têg, providing a much shorter route to Penrhyn and destinations beyond. A level crossing was established before the bridge (at 2m 36 ch).

Seven chains before reaching Pont Croesor road crossing a 110 ft siding, its point facing Portmadoc, ran off tangentially on the north side of the track. There was no road link, only

Figure 3.7

The original bridge at what is now Pont Croesor, photographed in the Croesor Tramway period. Note the wooden tramway fencing behind the group of people and the narrow, wooden-surface road way.

The bridge was completed in 1864 but significantly re-built when the railway was upgraded in 1922/3 in preparation for the arrival of the new Welsh Highland.

WHRHG collection

Figure 3.8

One of the original slab-built flood relief culverts under the Tramway embankment to the north of Pont Croesor.

David Allan 1985

a connection by public footpath, so access was problematical. However, one of the recorded users was the Hafod y Llan Slate Co.

Up to Pont Croesor and beyond there were a number of slab-built relief culverts as the Afon Glaslyn was tidal up to this point and the area was prone to flooding.

The line followed the road on a low embankment crossing a farm access road on the level then ran on to another access road at 3 m 33 ch (Ynysfor in WHR days), which it again crossed on the level with the gates across the tramway. The latter was a common feature along the route. There was a 40 ft siding, used for farm materials and manures, at this point.

The site of what was later to become Croesor Junction was reached at 3 m 60 ch. When the Welsh Highland Railway was completed from the original NWNGR terminus near the

Figure 3.9

Ynysfor in the 1920's, looking north towards Croesor Junction.

The siding and platform lay to the south of the level crossing visible by the more distant group of people.

C.L. Mowat M128

iBase 3194

village of Rhyd Ddu, the final section of the route to Portmadoc utilised refurbished Croesor Tramway metals to the south of the new junction. From this location, the quarries at the head of the Croesor Valley can be observed along with Cnicht and surrounding mountains.

Figure 3.10

Croesor Junction in the 1930's.

When the Welsh Highland was first established, the tramway made a direct junction close to the photographer's position. After the loop was installed here the tramway line was diverted to make connection with that loop, rather than with the main line.

H.M. Comber 1933

The tramway continued on a low embankment to Gareg-hylldrem with the Afon Croesor, a tributary of the Glaslyn, alongside and then to the old coast road and the level crossing at Pont Gareg-hylldrem (4 m 27 ch). This marked the end-on junction between the non-statutory tramway and the section which became the Croesor & Portmadoc Railway. Although legally part of the Welsh Highland Railway from 1923, the short section between Croesor Junction and Gareg-hylldrem was not reconstructed to take locomotives and – like the tramway proper – remained horse worked to the end. On the Croesor side of the level crossing there was a double siding to the left with a shed used for warehousing. This facility was usually referred to as Gwernydd.

Figure 3.11

Croesor Tramway tracks running from Croesor Junction towards Gareg-hylldrem.

The summit of Cnicht is prominent towards the centre of image with Cwm Croesor to its right.

F.R. Lawrence 1940

Figure 3.12

The level crossing where Croesor Tramway metals once crossed the Llanfrothen Road near Pont Gareg-hylldrem

Bill Rear

Figure 3.13

The confluence of the Afon Croesor (left) and the Afon Maesgwm (right) near the level crossing by Pont Gareg-hylldrem.

The tramway curved in from the right to join the line of the wire fence in the distance.

R.K. Cope 1949
via Roger Carpenter

The confluence of the Afon Croesor and the Afon Maesgwm lies just behind the site of this building and the tramway continued towards the quarries following the channel of the Maesgwm with the river to the left.

After a short distance it turned to the left crossing the Maesgwm to enter the loop and storage sidings at the foot of the Lower Parc incline (4 m 52 ch). A branch led off the right-hand loop line taking tramway metals to the inclines that led to the Parc (later Park) Quarry. The branch initially ran along the north-west bank of the Afon Maesgwm

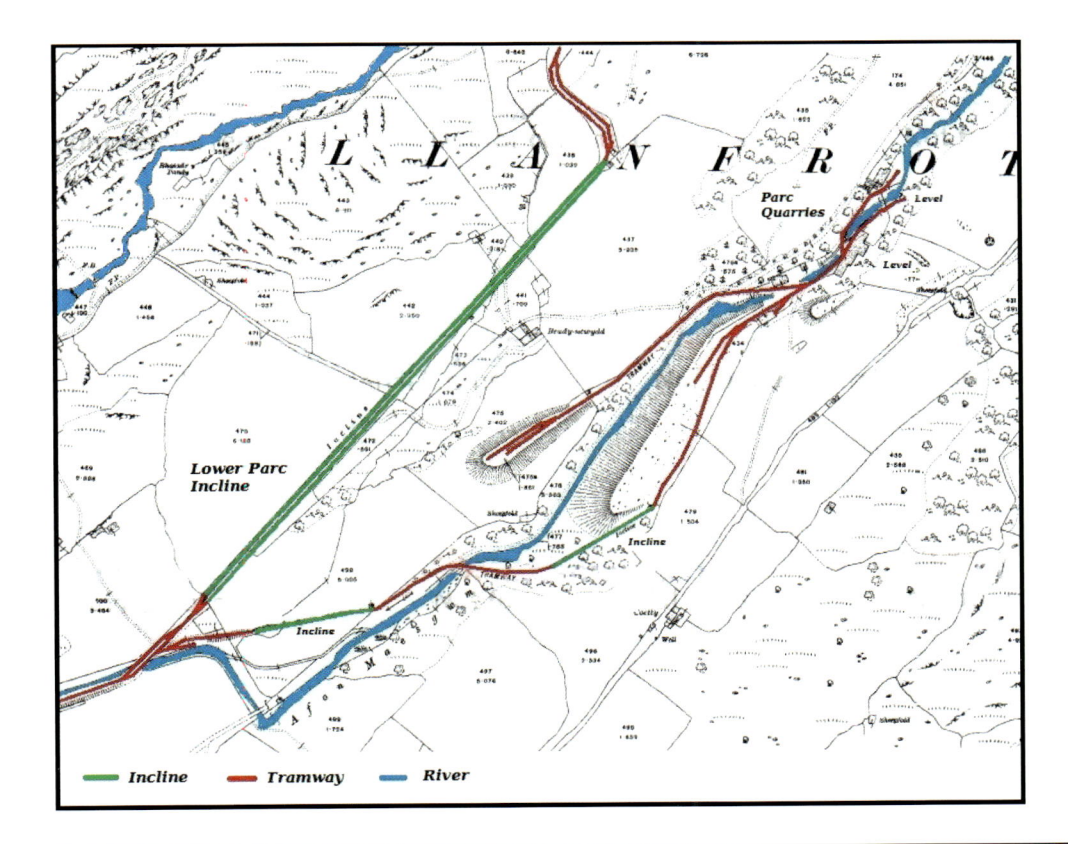

Figure 3.14

Trackwork at the foot of the Lower Parc Incline.

Note the turnout from the right-hand loop, roughly level with the lad leaning on the slate wagon. This junction gave access to the branch leading to the Parc Quarry.

C.L. Mowat M134

iBase 3191

and rose through two short inclines to reach the modest quarry workings. The branch track crossed the Maesgwm as it ran between the two inclines.

Figure 3.15

A map showing the layout of Parc Quarry access relative to the Croesor Tramway (Lower Parc Incline) and the Afon Maesgwm in who's valley the quarry was located.

Wikimedia Commons

From the inclines to Croesor village

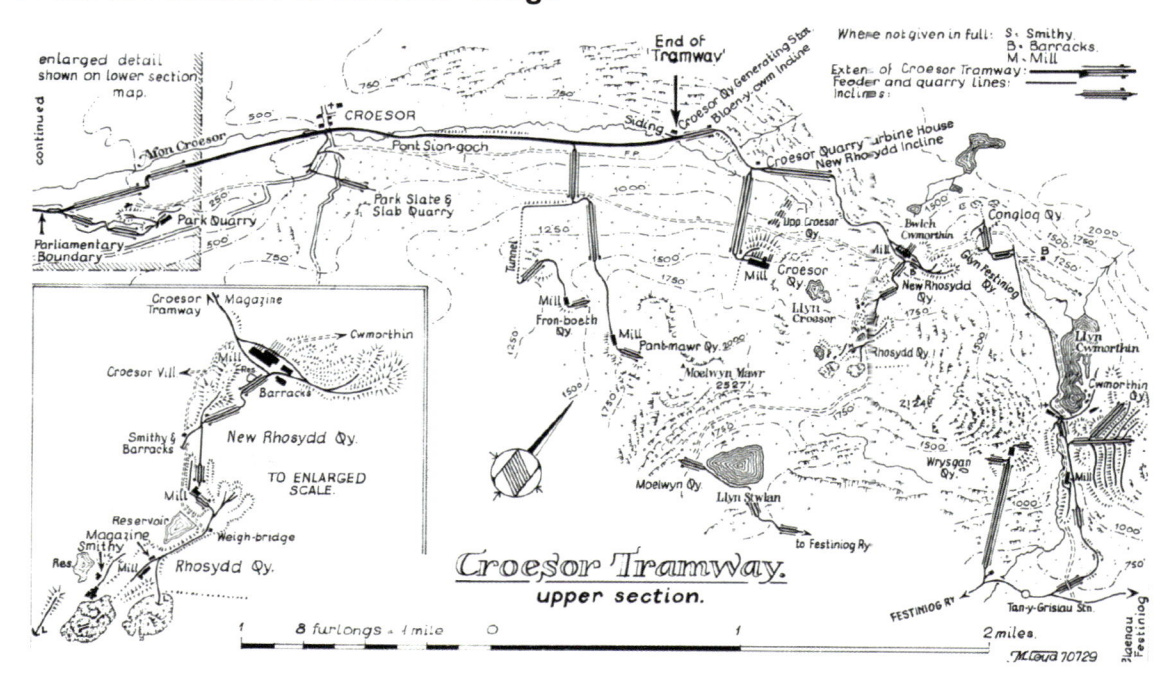

Figure 3.16

The route of the Croesor Tramway from the Parc inclines to the quarries in Cwm Croesor.

J.M. Lloyd

On the main tramway, the first incline, known as the Lower Parc incline, was almost 1,850 feet long, rising some 240 ft. At the top there was an S-bend to the left, with a loop, leading to the foot of the Upper Parc incline situated about 400 feet from, and about 25 feet above, the Lower Parc winding house. The Upper Parc incline was 1250 feet long, rising 130 feet. The Upper Parc drum house is now part of a private dwelling and can be seen from the minor road leading to Croesor village. The tramway has now risen from 19 ft above sea level to 290 feet at the top of the Lower Parc incline to 440 feet at the top of the Upper incline. Two separate inclines were adopted to reduce the length of each lift up the valley.

Figure 3.17

The remains of the Upper Parc incline winding house, looking up the slope of the incline with Cnicht framed in the opening.

Tudor Rose 1953

From the top of the Upper Parc incline the tramway ran north east, along a considerable dry-stone built embankment and, on approaching Croesor village, about 120 yards before the level crossing, there was another loop and siding on the right to serve the Park Slate and Slab Quarry, an undertaking distinct from the Park Slate Quarry. The branch served various parts of the quarry and one line crossed the main road leading to the village to serve various workings. A short incline gave access to the higher level parts of the workings. The quarry closed before the First World War.

Figure 3.18

The stone built embankment, described as 'The Croesor Railway Cei Mawr' by Moses Kellow in a letter to O.D. Jones in February 1907 - see Appendix 2.

This view looks towards Croesor Village from the occupation bridge giving access through the embankment near the head of the Upper Parc Incline.

D. Mitchell

On approaching the village, the tramway crossed the Afon Croesor on a stone pier bridge with slabs on top (a very old form of structure commonly known as a Clapper Bridge) and ran to an un-gated road crossing (5 m 64 ch). The tramway level crossing in Croesor village is adjacent to the now-closed village school. The school bunker was arranged so that coal could be delivered through a hatch in the wall direct from wagons on the tramway. Evidence of this access is still visible.

Figure 3.19

The clapper bridge over the Afon Croesor just before the tramway reaches the village.

D. Southern

At the same point, there was for a time a shed at which parcels and other goods could await collection and from which supplies could be delivered to Siop Goch, the village store just up the road. Further local supplies and deliveries, including for Moses Kellow at Bryn, were unloaded at Bryn-hyfryd, a little further up Cwm Croesor.

Figure 3.20

Photographs of horse-working on the tramway are rare. In this postcard from c.1920 we see an 'up' train standing at the level crossing in Croesor Village, the horse waiting patiently for business to be concluded. Note the small stone building to the rear of the train where the 'coal hole' was located and the shed alongside, standing on what is now the road through the village. Here the train driver would store goods for later pick up by the locals. Llys Helen is clearly visible to the right beyond the horse.

Figure 3.21

Despite its poor quality, this image from an earlier period does add to our knowledge.

Note that the building used as a coal-hole was much less developed and there was no sign of the shed seen in 3.20.

Again we see a stationary 'up' train being, at least in part, unloaded.

Both 3.20 and 21 are courtesy of Edgar Parry Williams

Cwm Croesor

North from Croesor village, the tramway crossed the Afon Croesor by a low slate built clapper bridge, Pont Sion-goch (6 m 00 ch), and ran along the bottom of the steep-sided Croesor valley, which is dominated on the left by Cnicht (2,265 ft). From this point the tramway was almost straight with an occasional embankment and cutting for some 1½ miles with a siding at Bryn-hyfryd for agricultural traffic (6 m 11 ch).

Figure 3.22

Above the village, Cwm Croesor is a lonely place and the tramway ran more or less unimpeded towards the next incline at Blaen-y-cwm.

The track in this photo is pointing directly at the winding house at the head of this next incline, visible towards the centre of the image.

John Keylock collection.

Further up the valley there was a junction to the right leading to an incline serving the Pant-mawr Quarry (6 m 60 ch). This incline originally followed the Parc layout with a two-stage lift separated by a level S-bend track layout at the mid-point. The lower section of incline was 1330 ft long, rising just over 500 feet whilst the upper section was just over 1500 feet long rising some 560 feet. The upper winding house was a little over 1600 feet above sea level.

When additional workings at Fron-boeth were developed from the late 1880s, the incline was re-laid as a single-lift to a point just above the original lower winding house, giving access to a tramway that ran southwest along the contour to a new tunnel cut through the Braich y Parc ridge to reach the Fron-boeth workings in Cwm Maesgwm. Pant-mawr was subsequently accessed via Fron-boeth. Now the incline was 1,450 feet long, rising 570 feet.

Figure 3.23

The remains of the Fron-boeth winding house with the quarry's tramway track bed leading off to the left.

The main tramway can be seen running from left to right across the centre of the image, with the site of the incline junction visible at the left.

Roger Kidner 1938

The tramway to the tunnel entrance was some 14½ chains in length, virtually level at an altitude of 1,140 feet, and the tunnel was 22 chains (480 yards) in length rising a little less than 60 feet on the run towards the quarry. On leaving the tunnel a very sharp left turn led to the foot of an incline which lifted the track to about 1,370 feet above sea level whence a short tramway of 16 chains led to the modest Fron-boeth working facilities.

The upper part of the original Pant-mawr incline was abandoned when this link to Fron-boeth was established.

From the foot of this collection of inclines, the tramway ran for another 32 chains beyond the junction to a final clapper bridge, which carried the tramway over the Afon Croesor for the last time (7 m 11 ch). This river crossing lay close by what would later be the site of the generating station at Blaencwm, built in 1904 to supply power to the Croesor Quarry. Two chains further on the line reached the foot of a 730 feet long incline which raised the track 190 feet up the end wall of the valley. The foot of this incline marked the end of the tramway and the start of quarry workings.

Figure 3.24

The final loops on the tramway before the Blaen-y-cwm incline. The electricity generating station stands to the left and the pipeline supplying water to the turbines can be seen on the left-hand side of the incline.

At the head of the incline the track turned right in front of the winding house.

J.F. Bolton 1940

Figure 3.25

Trackwork at the foot of Blaen-y-cwm Incline, looking back towards Croesor Village. Note the trackwork designed to handle wheels with double flanges. However, note that the turnout at the far end of the loop (Figure 3.24 above) catered for only single-flanged wheels.

J.F. Bolton 1940

North of Blaen-y-cwm

From the top of the Blaen-y-cwm incline the line turns to the right (east) and takes on a different character. It runs across the end of the valley, on well-built embankments and through cuttings, following the contours. The height here was some 820 feet above sea level. 14 chains beyond the winding house the track crossed the Afon Cwm-y-foel, a tributary of the Afon Croesor, on a high bridge. Following a collapse many years ago, the bridge was restored in the 1990s.

The Afon Cwm-y-foel was dammed higher up the mountain to create Llyn Cwm-y-foel, the lake that supplies the water to drive the Blaencwm power house turbines. The lake was at an altitude of almost 1500 ft, providing a head of over 850 ft.

Five chains beyond this bridge the line curved to the north to reach its upper terminus under the cliffs of Bwlch Cwm-orthin (at 7 m 41 ch from 'zero'). From here, two long and very steep inclines rise, one north-east to reach the Rhosydd Tramway and the other south-east to reach Croesor Quarry.

Figure 3.26

The final junction on the tramway. The incline to the Croesor Quarry is clearly visible running up towards the skyline. The foot of the Rhosydd Incline is seen in the lower left of the image.

This photo was taken when both quarries were operational - note the number of wagons in the Rhosydd 'queue'.

WHRHG collection

The Rhosydd incline lifted the track some 610 feet over its length of 1750 feet, to access the Rhosydd quarry tramway 1,430 feet above sea level. This tramway followed the contours in a generally eastern direction for half a mile to reach the Rhosydd Quarry mills. This tramway was particularly well engineered, running for much of its distance on high embankments hundreds of feet above Cwm Croesor.

The Croesor incline was just over 2,000 feet in length and raised the track 725 feet, but, unlike the Rhosydd incline, the Croesor winding house was situated in the quarry's mill area, avoiding the need for an intermediate high-level tramway.

Post closure

After the closure of the original Welsh Highland Railway, the stub of the tramway from Croesor village to the foot of the Blaen-y-cwm incline continued in use to carry agricultural products for local farms, until the late 1950s.

Figure 3.27

The remains of the tramway just above Croesor village.

A 2-ton slate wagon and the wooden body from a box wagon lie abandoned.

J. Peden 1960
Photo copyright Industrial Railway Society

The section of the Croesor tramway that ran from Croesor Junction to Portmadoc has been rebuilt as part of the reincarnation of the Welsh Highland Railway from Caernarfon to Porthmadog. The section between Croesor Junction and Pont Croesor opened to regular passenger service from 22nd May 2010. The section from Pont Croesor to Porthmadog reopened on 8th January 2011. Through services between Porthmadog and Caernarfon commenced on 19th February 2011.

Figure 3.28

F&WHR NGG16 Garratt No. 143 with a train from Caernarfon to Porthmadog approaches the new bridge giving access to Portreuddyn Farm, following the course of the original Croesor Tramway.

The Croesor Valley is visible beyond, nestling between Cnicht on the left and Moelwyn Mawr.

Martin Pritchard 2012

Chapter 4

Operations

On the upper section of the tramway wagons were worked by gravity down the inclines, which were of conventional balance type with overhead drums; the Lower Parc Incline deviated slightly in that it had a horizontal sheave in place of the drum. All the main line inclines and those up to Rhosydd and Croesor quarries were double tracked. Those on the Parc branch were single-tracked with passing loops.

The main tramway east of Croesor Junction always relied entirely on horse traction; memories vary as to whether a horse was used over ground at Ceunant Parc between the mills and the first incline and between the two inclines.

Prior to the advent of the Welsh Highland Railway in 1923, trains were always horse-drawn from the foot of the Rhosydd and Croesor inclines to Portmadoc. With the advent of the Welsh Highland Railway, trains continued to be horse-drawn to Gareg-hylldrem and normally through to Croesor Junction. It is remembered that, in the 1920s at least, the same horse worked a run right through from the quarry inclines to Gareg-hylldrem and vice versa.

On the flat sections worked by horses, operations were carried out under contracts with farmers and hauliers arranged by the Rhosydd and the Park and Croesor quarries. On the lower section the same arrangement applied until 1923 when the WHR opened and became responsible for all traffic between Croesor Junction and Portmadoc. The Croesor to Croesor Junction section in WHR days was horse worked.

Horses would work in pairs with a neck and collar and tug hooks linked to chain traces on each side to join a ring to which a towing chain was attached. The chain was attached to the wagons. The horses walked in the 2 foot space between the rails. They were hired from local farmers and in 1898 the norm was four wagons on a run, sometimes increasing to seven, with a morning and afternoon run to Portmadoc.

In *Y Cryman Medi* (*The September Scythe*) [1], the author, W.I. Thomas, describes his life from boyhood to minister of religion and records how in the early 1920s he worked on the Tramway. His work for two and a half years was following the horses that pulled the empty slate wagons. This involved hooking-up at the bottom of the inclines where the wagons were pulled up, or 'crewled up', to use the local term. The 'criwliwr' was John Williams; he was also a deacon at the old Ramoth Baptist chapel and he approached both jobs with equal seriousness.

> "If you ever thought of touching the brake on the incline you were really treading on sacred territory and he became most upset. Fortunately there were no accidents while I was working there, but I had a couple of narrow escapes. The brake on one of the wagons got bent and five loads started to run off, but luckily they were caught in time. At another time an incline rope failed; I was standing on the wagons but managed to jump off in time.

---

1 *Y Cryman Medi* ('The September Scythe'), Calvinistic Methodist Bookshop, Caernarfon, 1975

"Following the 'Run' was an interesting job. Once the horse has become accustomed to walking between the rails she was no trouble, and I was able then to ride in the wagon. I always carried a book with me, and I memorised a lot of poetry at that time. Coal and other merchandise were brought to the district by the 'Run'. I used to turn the wagons off into the siding, and later I would come back with the cart and deliver the coal and merchandise (flour etc.) to whoever required it.

"After two and a half years it was felt that it would be better to employ a younger man on the horses; the job was not really a full-time occupation. I sought work in the Croesor quarry, and the family with whom I was staying said that I could remain with them until I found some lodgings. I was promised work in the quarry as a 'llwythwr' (loader).

"I spent the next four years and two months at the quarry. It was a very happy and important part of my life, and I look back on it with pleasure. The work was hard, although I was given the freedom to do as I wished, and as we finished at 4.0 pm it left me with quite a bit of time to follow my own interests. It was a very close community of men; it did me a lot of good and I made many friends. There were only about eighty men working there, about half of them below ground as rockmen and loaders and the rest in the shed splitting and shaping the slates. My job was loading the slates, in the company of the slate clerk, and getting the loaded wagons to the top of the incline before they were run down. In the earlier days I had hooked at the bottom but now it was my job to lower the wagons down; luckily no accident ever occurred.

"Eight shillings and fourpence was the daily wage for a labourer at that time, and a quarryman received nine shillings and a penny; the loader came into the latter category. During hot summers the quarry lakes used to dry up; as there was no power to drive the machinery at the mill there was no work for the quarrymen, and they were laid off."

An essay written in 1961 by John Roberts of Penrallt, Croesor noted that:

"There was a pass-by after crossing the road near Tanlan with a wooden shed to keep goods, etc, since the tramway carried things other than slate, such as for Siop Tanlan [2]. The area depended on the tramway to bring in essentials for the homes and the farms, coal for the school and for the village. There was a pass-by near the school where these goods were kept, and the horses would have five minutes here. The tramway was very good (sic) with accidents. I remember my mother telling me of a girl who lived at Croesor Bach who tried to jump a wagon and missed; her leg went under a wheel and was broken."

In an article, *The Tramway Days*, the author, Robert Annwyl Williams [3], noted that in 1895, Croesor Quarry was re-opened under Moses Kellow and that:

"there were many loads carried up from Port, machinery and coal to re-work the furnaces, etc., before the re-opening ..... (and) continued until the arrival of

2 Siop - Shop
3 From Adrian Barrells' research

electricity. Hundreds of tons of sand and cement came up the tramway to build the dam [4]. Then 12 mules to carry it across the upper part of the valley to Cwmfoel, and then build a power station to supply electricity, using the stonework from the original workers' cottages that existed on the site (at Blaen Cwm) ..... there were visiting engineers here for long periods then.

"Following this work incoming traffic to the quarry was light, but as the local population grew so did their needs ..... there was no other way to bring in goods, etc., for a population of 200 or perhaps more. ....the first Sunday School trip from Croesor to Criccieth ... two or more wagons were hired and a horse to pull the train to Portmadoc, and then on the train to Criccieth."

Figure 4.1

Class 4-50/I built by the Pittsburgh Model Engine Co. (Works number 49604) under sub-contract for Baldwin, acquired by the FR in February 1925.

This petrol-engined locomotive on occasion worked 'bottom shunter' turns to Croesor Junction, picking up traffic from, and delivering goods to, the tramway.

S.W. Baker 1934.

As for the use of steam locomotives at Croesor, there has been considerable speculation about steam rail traction at Croesor, probably stemming from the quarry's use of portable or semi-portable steam engines, described (e.g. in the sale of Croesor United in 1874) as "locomotive steam engines". *The Pocket Book F* of the Industrial Railway Society postulated these to be two vertical-boilered 0-4-0 de Wintons. It was suggested that both had been supplied new to the quarry and that one of them was sold to the Coedmadoc Slate Company at Nantlle in circa 1898. At the moment there is no evidence for the existence of either of these locomotives, or indeed any other; it is probable that in pre-electrification days the *lefel fawr* was worked by horse.

The first documented use of powered traction was Moses Kellow's electric locomotive, built as part of Croesor's 1904 electrification scheme and this is described in Chapter 5, as is the use of diesel locomotives in the 1940s.

4 Llyn Cwm-y-foel

The Trains

The quarries kept very detailed records of the type and sizes of slates that were sent out. Analysis of the records [5] reveals the makeup of the trains and the loads carried. A sample page from the analysis is shown in Appendix 1 to this work.

Runs and Wagon numbers

The usual number of wagons in a run was four, rising from time to time to five, six or even seven. It was more usual for heavy traffic to be divided into morning and afternoon runs. Approximately 30 wagons were used for the Croesor Quarry traffic and the correspondence described in the later part of this chapter show that on the whole, there was never enough wagons.

Loads

In each way bill entry the size of slate in the wagon is followed by the quality; "b" = best, "m" = medium, "s" = second and "t" = third (damp course slates). Mediums predominate; the most popular size is 20ins x 10ins.

Order numbers

These are an enigma and the number 2087 appears regularly throughout the list. It probably designated a load which was merely for stock rather than for a particular customer.

Destination

"Croesor Siding" predominates and refers to the Croesor Tramway siding and wharf just north of the Cambrian Railways crossing in Portmadoc known locally as Beddgelert Siding. Slates were unloaded here for storage to await an order, transhipped for onward rail carriage, or run down to the Harbour to be loaded for transport by sea.

The section of tramway through Cwm Croesor also carried Rhosydd traffic and slate and slab from Parc joined the main route at the foot of the Parc inclines. A few additional items made the journey to Portmadoc; casks, boxes and ropes are mentioned. Any machinery needing repair also took this route.

The return runs carried the requirements of all three quarries, plus supplies for Croesor village and the surrounding farms, and there was a daily run to and from Portmadoc for the Croesor and Parc quarrymen and perhaps also for the Rhosydd men. The men made a monthly contribution towards the cost of these runs and this is the subject of some correspondence described in the section 'Using the Tramway' that follows on page 43.

WHR - exchange of traffic at Croesor Junction
From the opening of the WHR in June 1923 until mid-1924:

There was no loop at Croesor Junction – simply a single turnout, facing Portmadoc. Col. Mount, in his report for the Ministry of Transport of 29th May 1923, had required that the *"existing balance weight point lever should be changed for locking by the key on the (train) staff, and a trap point, operated by weighted lever, introduced on the tramway (now to be worked as a siding) to safeguard vehicles drawn by horses thereon fouling the main line."* The Railway assured the MoT that the various requirements had been met so we must presume these changes were amongst them.

---
5 Croesor Quarry Waybills: March-August 1898 Chwarel Croesor

Note that Croesor Junction was in mid-section between Portmadoc and Beddgelert. While it would have been quite practicable for an Up (i.e. Dinas to Portmadoc) train to set back and pick up wagons left on the tramway, it would be impossible simply to detach wagons bound for the valley from a Down train *unless a horse were there to haul the wagons clear of the WHR running line so that the points could be set and locked for the main line before the Beddgelert train departed.* Alternatively, provided the Train Staff was at Portmadoc, a run of goods could be made from there to Croesor Junction (by the bottom shunter, for example). This would be straightforward for bringing out loaded wagons from the tramway, but in the opposite direction the need for a horse to be present to haul the wagons on to the tramway could only be avoided if rope shunting were used. We know of no suggestion that such workings were adopted here. It must therefore be assumed that exchange of traffic was effected at Croesor Junction with the presence and use of a horse.

The 9th July 1923 working timetable [6] shows a trip from Portmadoc New to Croesor Junction and return at 0715, Mondays to Saturdays. The 1st October 1923 working timetable [7] shows the trip retimed to 1000 for October, and to 1225 from November onwards (although passenger services were drastically reduced before that service came into operation). There is no suggestion in either timetable of passing passenger trains being allowed time for shunting at Croesor Junction.

From mid-1924 until WHR entered receivership (5th March 1927)

The rigidity imposed by the long section Portmadoc – Beddgelert (and perhaps also the unwieldy nature of the working at Croesor Junction?) led the WHR to install a loop at the junction.

As noted on the sketch in Boyd, Vol. 2, [8] the initial proposal, for the loop be placed separately from (and south of) the tramway junction, was rejected by the Ministry in April 1924. It was replaced by a layout whereby the tramway connected with the loop which had been installed to the east of the running line.

On 3rd June 1924 Eric H R Nicholls confirmed to the Ministry of Transport that the sections of line between Portmadoc New and Croesor Junction, and between Croesor Junction and Beddgelert would be worked – as two separate sections – under Train Staff & Ticket regulations, thus signifying bringing in to use of the loop at Croesor Junction at about this date. Thereafter, traffic from the valley could be picked up by Up trains setting back on to the tramway, while traffic for the tramway could safely be detached from Down trains and left in the loop to be retrieved by horse (before the next movement that required the use of that loop, of course!). Similarly, runs of goods from Portmadoc could – holding the Portmadoc – Croesor Junction Train Staff and with some shunting – deliver and pick up traffic for and from the tramway, leaving the loop clear on departure.

The 14th July 1924 working timetable [9] shows a round trip from Portmadoc Old at 1010 Saturdays excepted. It also shows crossing movements at Croesor Junction (although, again, there is no suggestion of passing passenger trains being allowed time for shunting

---

6   WHH 20
7   TNA RAIL 981/513 (it appeared in the first edition (only) of Johnson's Illustrated History of WHR)
8   Boyd 1988 p.15
9   TNA RAIL 1057/2846/6 and WHH 24

there). This was the last working timetable issued so there is very little subsequent evidence about freight workings, although we know from Robert Evans' letter of 19th September 1925 [10] that the trip was then supposed to leave Portmadoc at 1215 (and be back at Old station by 1330).

During this year the trial of internal combustion locomotives – including, perhaps, the bottom shunter role? – took place; there is, however, no very self-evident reason why the working method at Croesor Junction should have been altered in consequence – despite Boyd's assertion [11] it seems highly unlikely that any locomotive would have worked beyond the Llanfrothen level crossing or, possibly, even beyond Croesor Junction.

After WHR entered receivership (5th March 1927)

Previously, the FR and WHR were worked, essentially, as a single unit – albeit with costs and revenue attributed to the two elements of the system – so this shunt will have been part of the bottom shunter's role.

From 5th March 1927, the costs incurred by the WHR may well have had to be even more carefully controlled and segregated. However, the run-down of passenger services makes it yet more unlikely that Croesor valley traffic was ever dealt with on any *regular* basis by such passing trains, so we must assume that the bottom shunter continued to trip to and from Croesor Junction (Boyd [12] asserts it reduced to thrice weekly at some unspecified date) - until horse working between the tramway and Gelert Sidings was substituted. This change from the use of FR power must have happened before 1930 because of the evidence of a traffic statement [13] for the four months ended 30th April 1930, called for by the Investing Authorities. Amongst the carriage rates quoted there is an even more interesting foot note stating that these rates included a haulage charge of 1s 6d per ton being "the payment made to the horse haulier, for hauling traffic, by horse, between Croesor Junction and Gelert Siding Portmadoc, and from the Rhosydd and Park & Croesor Slate Quarries. No haulage at all being performed by the W H Rly herein".

The context in which the question was asked made clear that this was the totality of the Croesor traffic - it can be concluded that none passed towards Dinas, and that all that did move did so by horse to Gelert Siding. Another tabulation for the nine months ended September 1930 showed that traffic over the Croesor Junction - Portmadoc section had been 59 tons "goods", 84 tons coal and 2373 tons slate (the nature of the "general" traffic conveyed to the tramway was discussed in *WHH* 37, p. 5). It seems one must conclude that by 1930 all Croesor valley traffic (up and down) was horse worked north of Gelert Siding, certainly in winter, *probably* also in the tourist season (what happened then is not clear - haulage outpayments were still being made but it's not possible to adduce whether they were at a reduced rate per ton to reflect the shorter haulage that would have been applicable had the horse working been restricted to north of Croesor Junction). Any Croesor valley traffic surviving after the loss of the Rhosydd outwards slate traffic later in 1930 would have been inwards goods worked by horse through to/from Gelert Sidings. However, there is no mention of outpayments to Kellow (or other horse contractor) in figures provided by the Receiver for summer 1931 or summer 1933 – so either there was

---

10   WHH 26, p. 6
11   Boyd, 1987, p. 132
12   Boyd, 1988, p. 38
13   at Caernarfon archives under references XC2/33/37 and /58; see also WHH 58, p.12

no traffic, or Kellow was working it and charging users direct rather than through WHR account.

To obtain the train staff (*WHH* 51, p. 9) for each trip during the period of passenger train service, Moses Kellow's subcontractor would have had to make the round trip either before the Up train appeared, or after the Down train had passed Croesor Junction on its return to Dinas. Boyd's claim of the passenger round trip in this period calling at Croesor Junction to pick up and set down slate wagons seems specious - the time allowed at Portmadoc wasn't really generous enough to do other than run round the train, let alone engage in shunting both there and at Croesor Junction.

The line south from Beddgelert must have had a somewhat "nominal" goods service at times: we know, for example, that in 1928 goods trains ran between Dinas and Croesor Junction to exchange traffic for the lower end of the line (*WHH* 45, p. 7), a similar function being performed by the "on request" Up mixed train advertised three days a week during 1929 and until September 1930 (it was advertised in only one direction so there was no need to go beyond wherever it had traffic for - to save unnecessary mileage). With the Croesor valley traffic being worked by horse through to Gelert Sidings, the "on request" train would rarely have run south of Beddgelert, during winter.

Using the tramway

Like most companies of the time, the Park and Croesor Slate Quarry companies kept letter books containing manuscript copies or summaries of all letters and telegrams, with some printed or typescript copies. They were all written by Kellow, either on his own behalf or for his father. The letter books represent only one side of a correspondence; the files of letters from directors, suppliers and customers have vanished without trace.

The transcription of this information, by Adrian Barrell, was undertaken between 1998 and 2002. The Letters Books were badly affected by damp with many pages welded together creating significant gaps in the chronology. He tried to include not only items which were felt to be important but also the more trivial and sometimes the downright amusing.

The extracts (presented as Appendix 2 to this work) relate mainly to the Tramway although the bulk of the letters and the transcriptions, understandably, cover quarry issues. They run from 12th July 1881 to 18th March 1914 and provide a flavour of the frustrations and problems arising from use of the tramway and the dealings with, for example, the Cambrian Railways in Portmadoc. The correspondence reveals that problems faced by the quarries included a shortage of wagons and the supply of new ones. In December 1900, there is some correspondence about the use of dumb-buffer trucks by the Cambrian Railway and the damage to the slates carried in them. The transhipment of the slate from the Croesor wagons was also causing a problem.

Much of the traffic up the valley consisted of equipment for the quarries.

Not all the extracts refer to the tramway and from time to time more domestic matters intrude, such as the lack of harmonium lessons, complaints about slow payment, a broken bottle of whisky and the loss on a LNWR train of Moses Kellow's 7ft long muffler.

The immediacy and efficiency of the postal service between the quarry offices and, for example, Portmadoc is underlined by a letter on 19th May 1914 to Jones & Co., Portmadoc asking 'Will you kindly send 2 boxes of candles up to Croesor Quarry per tomorrow's run.'

Figure 4.2

A sample Kellow invoice to the WHR for 'haulage between Croesor Junction and Gwernydd'.

This invoice was eventually settled, but not for 3 months!

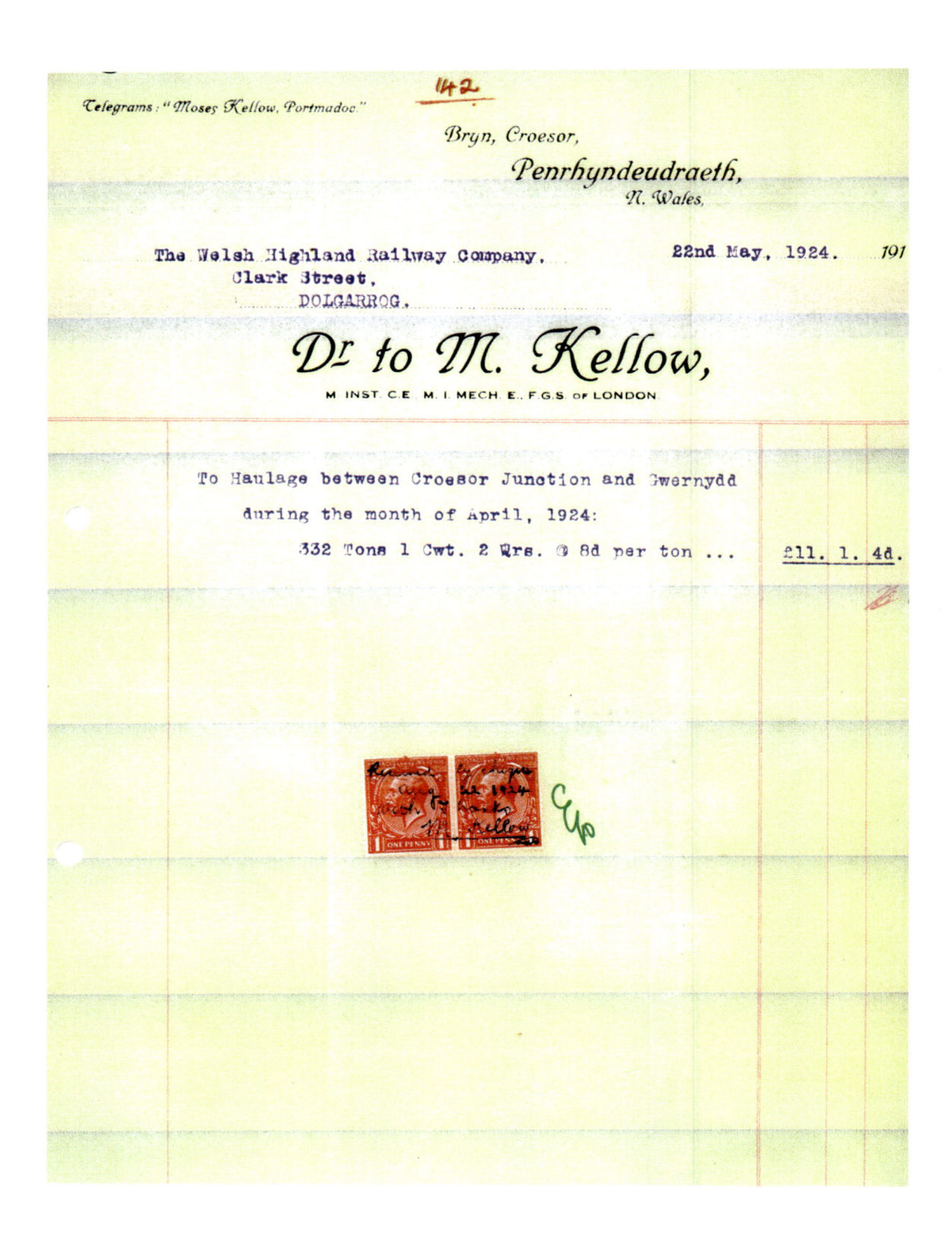

Chapter 5
Sources of Traffic

Background

A number of branches leading from the tramway served the quarries of Cwm Croesor. The first of these branched off at the foot of the Lower Parc Incline and served the Parc (later Park) Slate Quarry. Just south of Croesor village, the next branch connected with the separate Park Slate and Slab Quarry, curving on the eastern side of the tramway and reaching the quarry by a short incline.

Half way between Croesor village and the foot of the Blaen-y-cwm incline, an incline climbed south-eastwards up the side of the valley. At the head of the incline a tramway ran to the south west along the 1,150 ft. contour line of the hill to connect, via a tunnel and incline, with the Fron-boeth Quarry. Formerly, the incline was slightly shorter and gave access to a second stage incline to reach Pant-mawr Quarry. The Pant-mawr access was abandoned when Fron-boeth was developed.

At the head of the valley, are two immense inclines one climbing over 700 feet south south east to Croesor Quarry, the other heading north east and rising 650 feet to connect to a well-engineered ½-mile long tramway to Rhosydd Quarry. The latter incline forms a parabolic curve and at its head climbs at 1 in 0.97, one of the steepest inclines in Wales.

The goods outward from the Croesor valley were primarily slate products while inward goods were supplies for the quarries and Croesor Village including coal at a time when every home had a coal fire. Other inward goods included building materials, explosives, agricultural requirements and flour.

Traffic from Rhosydd and Croesor was primarily slates while Parc produced slates and slab. However, a wide variety of other traffic was carried including casks of paraffin, ropes and building materials and coal to Croesor quarry and Blaencwm power station.

The table below shows examples of the tonnages produced by various quarries and that would have been sent down the tramway.

Park Quarry		Croesor		Park & Croesor		Rhosydd	
Year	Tons	Year	Tons	Year	Tons	Year	Tons
1907	1762	1862	74	1907	2,465	1897	3,124
1911	3436	1863	555	1908	2,620	1898	2,727
		1864	1,189	1911	3,819	1907	1,561
		1865	946	1912	3,970	1908	2,009
		1873	1,611			1926	1,796
		1876	691			1927	1,598
		1877	884			1928	1,436
						1929	1,473
						1930	899

Table 5.1

A summary of quarry outputs.

A more comprehensive summary of Croesor output will be found at Appendix 3

By courtesy of David Payling

Not all the loaded wagons constituting an 'inward' run from Croesor Junction would need to go even up the first incline and there were therefore sidings at Gwernydd, close to the Nantmor to Llanfrothen road, where goods were unloaded, sometimes for temporary storage in a shed. Tan-lan, Parc Quarry and farms up the road towards Nantmor would be supplied from here.

The busiest section was between Beddgelert Siding and Portmadoc harbour, which included slate from the Festiniog Railway for transhipment to the Cambrian until Minffordd exchange sidings were opened in 1872. In later years, when Beddgelert Siding was busy, traffic from Parc and Croesor quarries was routed through to Minffordd exchange yard.

The Quarries

Rhosydd Quarry

Lying some eight miles north east of Portmadoc and two miles west of Blaenau, on the north side of the Moelwyn range, Rhosydd Quarry began as a small concern in the early 1830s with the first serious attempts at quarrying starting around 1852. However, it was stunted in its development by poor transport. For 10 years, it struggled on despite an injection of capital and only surged ahead when connected to the Croesor tramway.

The surface operations were at the col dividing Cwm Orthin and Cwm Croesor, Bwlch Cwm-orthin, some 1487 feet above sea level. The first workings were located half a mile to the south of the col at 1850 ft near the ridge that joins Moelwyn Mawr (2527 ft) and Moel yr Hydd (2124 ft).

Before the tramway was built and while the quarry was being developed, slate was moved by pack horse and the cost of moving the slate was very high. The original route in the 1860s from the highest workings is believed to have been round Llyn Trwst-y-llon (Llyn Stwlan), joining traffic from Moelwyn Quarry. Following the expansion of Rhosydd Quarry, the slate was taken by pack horse to the Cwm-orthin/Croesor track at Bwlch Cwm-orthin. From here, the slate could go either to Tan-y-grisiau and the Festiniog Railway and thence directly to Portmadoc, or via Cwm Croesor, by packhorse, and so to Portmadoc. In the quarry, there was an internal tramway of some 4½ half miles, part which was underground.

The Rhosydd Slate Company was formed in 1853, and became a limited company in 1856 with English directors and shareholders. Moving the slate to market was made more difficult by the attitude of the Cwm-orthin Slate Quarry, as the most obvious route to the Festiniog Railway lay through their site.   The solution was found in 1864 with the opening of the Croesor Tramway, to which the quarry was connected by one of the longest single-pitch inclines in Wales. Development work absorbed huge amounts of investment money but the company, unable to make adequate returns, went into voluntary liquidation in 1873.

The quarry was auctioned in 1874, and the New Rhosydd Slate Quarry Company Ltd was formed with Welsh directors and three-quarters of the shareholders were from the local area. The quarry prospered for a while, but profitability declined steadily. In 1900, a large section of the underground workings collapsed. New chambers were opened but sales were hampered by a slump in the slate industry and at the onset of the First World War the quarry was deemed "non-essential" and mothballed for the duration.

It reopened in 1919, but was in a poor financial position and in 1921 the company went into liquidation. The quarry was then bought by members of the Colman family, better known for producing mustard, who had bought Croes y Ddwy Afon Quarry the year before.

Figure 5.1

The main entrance to the 'new' Rhosydd Quarry was via an adit which accessed Level 9 of the underground workings.

Rhosydd Quarry Booklet

However, like many other quarries in the 1920s it continued to lose money as the product failed to sell sufficiently well. Roofing tiles were used more and more by the building industry after 1924, and where slates were used, they were in many cases cheap imports. Rhosydd closed on 13th September 1930 and was then mothballed, two caretakers being employed and the pumps kept going. Fuel oil in drums proved to be the ast traffic for Rhosydd and the caretakers' job was to maintain the track for this traffic. To bring the wagons of oil up the load was counterbalanced by loaded slate wagons of waste and this continued until 1944.

In 1947 it was bought together with its neighbours Croesor and Conglog by Captain J S Matthews who owned Craig-ddu and Manod quarries but plans to reopen came to nothing and in 1948 the pumps were turned off. The track on the incline and much of the machinery was then removed for scrap.

Figure 5.2

The main surface mill area at Rhosydd with Cwm Croesor falling away beyond.

Rhosydd Quarry Booklet

Nevertheless, Rhosydd was not entirely finished as a small syndicate of slate makers had permission to produce roofing slates from slabs extracted from the walls of the main mill. This lasted from 1949 until about 1954 and since then Rhosydd has been empty and at the mercy of the elements, sheep, ramblers and underground explorers. [1]

Rhosydd Tramway and Incline

On 1st October 1863, Hugh Beaver Roberts gave permission, to the owners of Rhosydd, to build a tramway and incline to connect to the Croesor tramway, in return for a royalty payment of 2d per ton.

Figure 5.3

The Rhosydd Tramway looking towards the incline head (left) and back towards the Quarry (right)

Dave Sallery

The tramway ran for about half a mile from the quarry stacking yard to the incline head, on a shelf cut into the mountain supported in some places on a dry stone wall, with a single track and a loop at the incline head. The incline falls over 600 ft and is one of the two steepest inclines in Wales and was gravity worked. It is 12 ft 4 in wide and was double tracked with 35 lb yard rail. The drum head for the cables was 50 ft above the incline due to the lack of space.

Figure 5.4

As space at the head of the Rhosydd incline was severely limited, there was no room for a 'traditional' winding house. Instead, the drum was set into the hillside some 50 feet above the level of the quarry tramway.

Dave Sallery

At the foot of the incline was a scissor crossing and a loop for marshalling wagons. Trains of five to six wagons were hauled by horses to Portmadoc where part of the Greaves wharf was leased. There was also a wharf at Beddgelert Siding. In 1928 the quarry owned 54 wagons.

The incline was worked by the drivers of the horses. Traffic varied from one wagon per day up to eighteen, with an average load of 1 ton 13 cwt. The bulk of the outward traffic was roofing slates, slab, scrap iron and wool from Cwmorthin sheep. Inward traffic included coal for the boilers of the steam engines used in pumping, for example 331 tons in 1867, coke, gunpowder, timber, glass, windows chains, ropes, wire, winches, iron pipes, food, oil and grease.

1 Lewis M J T & Denton J H, 1994

Croesor Slate Quarry

Quarrying at Croesor on the western edge of the Blaenau Festiniog area commenced in the 1840s with small scale underground workings and by 1861 there were two companies in operation. They amalgamated in 1865, a year after the quarry was connected to the newly opened Croesor Tramway. While a considerable investment was made in development work, the volumes of saleable slate produced were small, amounting to just 226 tons in 1868. A change of ownership in 1875 did little to improve the profitability of the quarry, and this first phase of working ended in about 1878 or just after.

Access to the underground workings was by a single half mile long adit, and the surface mill was powered by two water wheels. These were later supplanted by a large Pelton Wheel (an impulse-type water turbine) and a small amount of steam power. The only surviving water wheel was a large one beneath the fitting shop, driving the New Mill. [2]

Figure 5.5

A loaded slate train leaving the Croesor underground workings behind an electric locomotive - note the overhead wires.

Henry Hall - Inspector of Mines - 1904

In 1895 the quarry re-opened under Moses Kellow who improved the operation of the mine and output increased to between 5,000 and 6,000 tons per year with a workforce of some 300 men at its peak but later declined until closure in December 1930.

The Croesor Quarry was the largest connected to the Croesor Tramway, by an incline with a fall of over 700 ft. It was over 1600 ft above sea level and was one of the first quarries to use hydro electric power, with a reservoir, Llyn Cwn-y-foel, feeding a small power station at the foot of the Blaen-y-cwm incline.

The quarry was unusual in that it was largely underground. Some of the slate waste from new workings was used to back fill the worked out old chambers. However, there is a great deal of external or outbye waste at Croesor and The New Mills are built upon it.

The first documented powered traction was provided by an electric locomotive, built as part of Kellow's 1904 electrification scheme. Kellow believed this locomotive to be the first electric mining locomotive in the country. It may have been the first in Wales but it was by no means the first in Britain [3]. By the late 1880s there was a 22" gauge 15 hp

---

2 Adrian Barrell, August 2018 - note to Nick Booker
3 Adrian Barrell, August 2018 - note to Nick Booker

electric locomotive at work in the Lucy Tongue Level of the Greenside lead mine on the north-eastern slopes of Helvellyn in the English Lake District.

Figure 5.6

Kolben's electric locomotive at work on the Croesor Quarry mill level

Slate Trade Gazette Feb 1911

Kellow's locomotive was of 30 hp, using two 15 hp motors drawing 220 volts dc from a single overhead wire with running rail return. Kellow 'designed' it to haul 12 wagons, about 18 tons, at speeds of up to 8 mph; he said that it had successfully hauled twice this number. The general belief had been that the locomotive was Kellow's brain-child. In fact, the locomotive was designed and built by Kolben's of Prague and Kellow's only contribution was to provide basic information such as gauge, loading gauge, load to be hauled, coupling height, and so on. The evidence is to be found in the Letter Books of the Park and Croesor Slate Company [4].

Figure 5.7

Emil Kolben

Wikimedia Commons

Emil Kolben was born in 1862 and died in 1943 (in the Theresienstadt concentration camp) and was thus a contemporary of Moses Kellow. Kolben's were electrical manufacturers on a large scale, comparable to BTH or Metropolitan Vickers in Britain. They were responsible for many Czech power stations, and they designed and installed the electric street tramway system in Prague. Clearly a small electric locomotive posed

4 Barrell, 2002

no problem to their drawing office, and it was included in the complete package of alternator, dynamos, motors and control gear for the Croesor contract.

Kellow saw the locomotive for the first time when he re-visited Prague in the summer of 1903 to check the efficiency of the Croesor equipment. His only observation at that time was that there was insufficient headroom in the cab, found in a letter to Witting Eborall, Kolben's London agents, and contractors for the Croesor installation. The locomotive had been delivered by 30th November 1903.

Figure 5.8

One of Kolben's electric locomotives at his factory in Prague - it is believed it was under test prior to export to Tasmania.

Simon Darvill collection.

Croesor quarry was closed in 1930 with the sale and disposal of assets in May 1932, although between 1934 and 1939, it was kept under care and maintenance.

Summary of the quarry's output between 1895 and 1941 is presented in Appendix 3.

Figure 5.9

"The Last Load".

The final shipment of explosives leaves the Croesor Quarry

Photo courtesy of Ron Podmore via Martin Pritchard

In 1942 it was requisitioned by the Ministry of Supply for use as an explosives store and from 1949 was used for the same purpose by Cooke's Explosives of Penrhyndeudraeth, Gwaith Powdwr, where explosives had been manufactured since 1865. When Cooke, the owner, retired in 1958, the company was acquired by ICI and continued production until 1995. ICI had bought the quarry in 1964.

The explosives and propellants were stored in the quarry's various chambers and these were serviced by three 4-wheeled Ruston and Hornsby 20 HP diesel mechanical locomotives built in 1939 and 1941 and bought second hand. In 1971, the Central Electricity Generating Board, operators of the Ffestiniog pumped storage power station, became aware of the storage of explosives at the quarry and worked out that if an explosion took place it could damage either of their two dams. They therefore drained both reservoirs. Cooke's then began removing the explosives for dispatch at the rate of 250 tons per week to other ICI plants, via British Rail at Penrhyndeudraeth. The last load of explosives moved out in September 1976 and the quarry closed.

Blaencwm power station continued in operation after the quarry closed, supplying electricity to the national grid until the 1950s, when it too was closed. The dam on Llyn Cwm-y-foel was breached and much of the pipeline was removed. The building became an outdoor pursuits centre. In 1999, National Power Hydro refurbished it as part of a £1 million project, which included the provision of a new underground pipeline from the reservoir, and underground cables to connect it to the national grid. It is capable of generating 500 kW, and the work was carried out as part of the government's renewable energy programme.

In the early 1970s the quarry was purchased by the Ffestiniog Slate Company [5] with a view to reopening it as a working slate mine. However, planning permission was refused and in the later 1970s most of the remaining mine infrastructure was removed for use in the Oakeley Quarry, also owned by the Ffestiniog Slate Company

Pant-mawr

Pant-mawr Slate Quarry was opened in the 1840s and operated in conjunction with Moelwyn Slate Quarry on the east side of Moelwyn Mawr. A series of thirteen levels with their attendant spoil heaps form a line across the hillside. The chambers are accessed via adits on the 1500 ft contour. From 1853, slates were carried by pack animals down Cwm Maesgwm to the Festiniog Railway at Penrhyn. In 1863 a tramway was constructed across the south-west face of Moelwyn Mawr to the head of a double-pitch incline falling 1100 ft to Cwm Croesor where it joined the Croesor Tramway.

The mill, built of roughly-squared country rock and of which only fragments of its walls remain, was probably built in the 1850s on a terrace part excavated and part supported by a massive retaining wall. Two inclines connected the mill level with some of the adits at a higher level and to the 1863 tramway below. Pant-mawr amalgamated with Fron-boeth Quarry in 1886 and surface operations ceased.

Traffic was first noted in September 1877 lasting until May 1892 and the quarry closed in 1908.

Fron-boeth (originally Foel-boeth)

Fron-boeth emerged from an ambitious project to extend and develop the Pant-mawr quarry workings and opened in 1886. The chambers were below the Pant-mawr workings

---

5 wikipedia.org/wiki/Croesor_Quarry (no citation - accessed 10/10/2018)

and accessed by three adits serviced by an incline, which brought the slate down to a steam powered mill. Finished product was taken out by a horse worked tramway to the head of a long stone embanked incline. From the foot of this incline a tunnel almost 500 yards long ran from Cwm Maesgwm to Cwm Croesor through the Braich y Parc ridge. Entering Cwm Croesor, the line turned right, up the valley, to link with the head of the rebuilt Pant Mawr lower incline to make connection with the Croesor Tramway. The lower incline had been extended and the higher incline to Pant-mawr had been abandoned. Use of the Croesor Tramway commenced in 1891. Fron-boeth initially employed around 50 men with output declining by the early 1900s, work finally stopping by the outbreak of the First World War.

Berth-lwyd

A small pit working quarry, in the valley of the Nanmor, with two levels connected by inclines and a mill to process the slate. The quarry was active in the 1860s and 1870s employing 30 men with an output of 170 tons of slate per year. It is believed to have had cart access to the Croesor Tramway.

Cefn y Braich

Slate was quarried rather than mined, working at two levels on hillside terraces, at Cefn Y Braich. The upper incline from the Fron-boeth Mill passes through the Cefn y Braich waste tips. It was in use between 1877 and 1883 employing 20 men and with an out put of 240 tons per annum. The quarry later became part of the Pant-mawr and Fron-boeth quarries. Output was sent out via the Croesor Tramway.

Parc Quarry

The Parc Quarry, located to the east of the lower Parc incline, opened around 1866 and was also known as Ceunant Parc, to distinguish it from the Parc Slab Quarry, a little further up the valley. Access was via an adit, from which chambers were developed with material processed at a water-powered mill. As the product was unsuitable for roofing slates most of its early output was slate slabs.

In 1883, there were 15 men working the quarry producing 351 tons of finished slate. The quarry was connected to the Croesor Tramway at the foot of the Lower Parc incline. To reach that point, a tramway along the east bank of the Afon Maesgwm descended from the mill via a short incline and then crossed the river by a bridge and descended via a second incline to reach the junction. The quarry started using the Croesor Tramway around 1872. The quarry became part of the Park & Croesor Slate Quarries Co Ltd in 1895. The sidings at the bottom of the first incline by the Croesor Tramway were known as the Moses Kellow sidings.

Work ceased at Parc around 1920 and Croesor in 1930 although spasmodic working continued with occasional trains, along the tramway, often only of one wagon, sent out to the standard gauge railway at Gelert Siding.

Parc Slab Quarry

Further up the valley from the main Parc Quarry, this was small working pit with a sawing mill and operated in conjunction with the Croesor Quarry. The finished product was taken away on a branch of the Croesor Tramway, which joined the mainline near Croesor Village.

Other locations served by the Croesor Tramway

Pont Gareg-hylldrem

Just after the tramway crossed the Llanfrothen road on the level near Pont Gareg-hylldrem (at 4 m 27 ch) a goods station, known as Gwernydd, was installed. A large storage shed was served by two short sidings off the tramway with their points facing Portmadoc. The shed was located close to the confluence of the Croesor and Maesgwm (Figure 3.13). This facility served the local farms and communities, such as Erw-fawr and Tan-lan.

Beddgelert Siding

Beddgelert Siding (later Gelert Siding) was the transhipment facility for slate from the narrow gauge to the standard gauge north of Portmadoc between the Aberystwith & Welsh Coast Railway (later Cambrian Railways and GWR) and the Croesor Tramway, later the WHR.

The siding gained its original name as part of a scheme for a standard gauge branch to Beddgelert. Work had begun in 1864 and had progressed towards Prenteg, in part using the pre-existing Creassy Embankment. This had been created in the early 19th century to gain land from the Glaslyn estuary, prior to the construction of Maddock's great embankment, the Cob, that was later crucial to the Festiniog Railway's approach to Portmadoc. However, construction of this line ceased, having got no closer to Beddgelert than the Croesor Tramway had managed about one year earlier. The site was renamed Gelert Siding after the opening of the WHR in 1923, presumably to avoid confusion with sidings actually located at Beddgelert station.

The AWCR opened to traffic in October 1867 and Beddgelert Siding opened around the same time to traffic for slate as well as general goods such as coal, flour from the mill in Portmadoc, agricultural products for farmers and outward supplies to the quarries in the Croesor valley.

The standard gauge siding began as the up running line in Portmadoc station, effectively as an extension of the loop. Trains to Dovey Junction would have left the loop by means of a crossover. Beyond this the straight line became Beddgelert Siding proper, and within a few yards a reverse shunting neck gave access to the cattle dock on the site of what is now the Welsh Highland Heritage Railway car park, opposite the Queen's Hotel. The siding curved away in a north easterly direction, enclosing the area now known as Gelert's Farm within the third side of a triangle of railway lines (standard and narrow) before terminating near Pen-y-mount.

The Croesor Tramway ran alongside its east side, serving transhipment wharfs and stacking grounds for the storage of slates. There was a loop off the tramway's main line so shunting could take place either to the slate wharf or the two sidings by the cattle dock. There was also a transhipment shed [6], for items such as flour, and a weighbridge. In the early days, the Festiniog Railway also provided traffic, via reversal in the harbour,

---

6 Correspondence and plan (dated 1919) in the National Archive (Kew) (Rail 1957/ 1962)

until the exchange sidings at Minffordd were built. Nevertheless, Gelert siding was still used thereafter by the FR at busy times.

The narrow gauge sidings were partly relaid in 1922 with a revised layout as part of the building of the Welsh Highland Railway. In the 1929-1934 period the goods loop was used by the locomotives to run round trains, when WHR trains terminated north of the Cambrian line crossing as an economy measure. They also did so on occasion subsequently.

The Croesor section of the WHR was left in place after the remainder was dismantled in 1942, to allow for a possible resumption of quarrying in the Croesor valley after the war. It was removed in 1949 when it was clear such a revival was unlikely.

The site of the standard-gauge siding subsequently became the home of what is now the Welsh Highland Heritage Railway which purchased it from British Rail in 1975. Soon afterwards they were able to purchase the adjacent Gelert's Farm to use as an operational base. Since 1980, WHHR passenger trains have run along the site of the siding, terminating at Pen-y-mount. Between March 2007 and November 2008, trains ran beyond this on to the original Croesor/WHR trackbed to Traeth Mawr where there was a temporary loop.

Moel-y-gest Granite Quarry Portmadoc

The quarry was situated on the northern slopes of Moel-y-gest with a rope worked incline to the foot of the mountain for the extraction of granite and opened in the mid 1870s. The quarry was served by the Cambrian Railways, later the GWR. In 1903 a narrow gauge line was established that ran via the Gorseddau Railway with running powers over the Croesor Tramway to Portmadoc Harbour and the Festiniog Railway.

This siding, as far Portmadoc station goods yard, may have been of mixed gauge. The line was closed in 1907 but reopened with a standard gauge connection to the Great Western Railway in 1919. The quarry was closed and dismantled by 1950.

The remains are spectacular and dominate the local landscape of Porthmadog and Tremadog [7], as a massive linear area of extraction along the northern slopes of Moel-y-Gest, centrally accessed by a steep incline down to the valley floor. On site [8] the foundations for long scrapped machinery remain, the walls of the drum house and a narrow gauge wagon turntable.

Portmadoc Flour Mill

The Portmadoc Flour Mills [9] building on the corner of Snowdon Street, Portmadoc was erected in 1862 as a steam-powered roller mill. Production continued through the 19th century until the mid 1940s and the building was then converted to other uses including a studio pottery from the late 1970s until the mid 1990s. Lately the building has been the subject of various projects for conversion into apartments.

The Croesor Tramway and later the Welsh Highland Railway passed the boundary of the mill and were used to send bags of flour up to Beddgelert Siding and beyond and sacks

---

7   www.coflein.gov.uk/en/site/408992/details/moel-y-gest-slate-quarries (accessed 10/10/2018)
8   https://www.aditnow.co.uk/Mines/Moel-Y-Gest-Granite-Quarry_1252/ (accessed 11/10/2018)
9   https://tinyurl.com/y8gfmrq9 (accessed 11/10/2018)

Figure 5.10

The flour mill loading dock in the Welsh Highland period. Note how the roof has been cut back to clear the WHR main line that curves to the right towards Portmadoc New.

The line to the left provides siding access to the Welsh Slate Company's mill and follows the start of the Gorseddau Railway route towards Tremadoc.

Ken Hartley 1930

of grain from the ships in Portmadoc Harbour, and possibly Beddgelert Siding, to the mill. Flour was also sent via the Festiniog Railway to the bakery at Penrhyn up until 1946 [10].

Adjacent to the mill there was a 'loading shed' [11] covering a loop of the narrow gauge railway, that was used for loading the railway wagons under cover (see Figure 3.4). Whilst the Croesor Tramway remained horse drawn, it was possible to have an all-over roof on this shed, covering both of the double tracks at this point, but with the introduction of steam locomotives with the opening of the WHR in 1923, the former arrangement could not continue and the 'shed' was rebuilt as an awning only covering the east loop.

Richard Williams Slate Works / North Wales Slate Co Ltd

Figure 5.11

The Welsh Slate Company's works photographed just beyond the turnout seen in Figure 5.10

The trackwork is the surviving slate works sidings - the Gorseddau main line ran to the right of these, along what by the time this photo was taken had become a footpath.

R.K. Cope 1949
via Roger Carpenter

The slate mill was served by a siding off the west side of the flour mill loop and followed the formation of the Gorseddau Junction and Portmadoc Railway Co (Gorseddau Tramway). As late as 1943, there was a request for a train service to their private siding. The request arose from petrol rationing as they had not used rail transport for some years.

---

10 www.festipedia.org.uk/wiki/Penrhyn_Station (accessed 11/10/2018)
11 Berks and Jones, 2009

They received short shrift from the railway company. In the past the slate works had paid the Festiniog 15 shillings a year for having slab brought there from Blaenau. Theirs was the last operational siding off the WHR [12].

Glaslyn Foundry

Glaslyn Foundry was located on the site now occupied by Wilko, close to the point where the Croesor Tramway and later the Welsh Highland Railway crossed Portmadoc High Street. The foundry opened in 1848 and produced a wide range of products including slate working machinery, doors, tools, cranes and railway equipment. The site had a siding linked to the Croesor Tramway and protected by a catch point. The foundry closed in 1981.

---

[12] Boyd JIC, 1989

Chapter 6

Personalities

Moses Kellow (c1862-1943)

Moses Kellow, JP, MInstCE, MIMechE, MIQ, FGS was born in 1862 at Delabole, Cornwall, close to one of the largest slate quarries in Britain. His father worked at Delabole Quarry. His parents moved to Carnarvon when he was some three years old. He died in 1943.

He was Manager, and with his father part-owner, of the Croesor Quarry from the mid 1890s until closure in 1930 and beyond. He was one of the most remarkable characters in the history of slate mining in the area. He did much to change old quarrying practices. His two major, but by no means only, achievements were the introduction of electricity into those quarries he managed and the design and build of one of the most powerful portable rock drills ever made.

The nature of Kellow's innovative imagination is illustrated in a school exercise book he used as a note book. [1] The first few pages contain his childish drawings but the latter part contains his thoughts on the hydro electrification of Croesor Quarry. The notes are dated January 1902 but he had obtained quotations for dynamos as early as May 1900.

1 Barrell, 2002

Clearly he was delving into what for him was new field but apart from his subsequent progression from direct current to three phase alternating current and far greater power these notes contain in embryo the whole of the scheme in a long list of potential projects including providing an electric pump to supply hydraulic power for his Keldrils. The only listed proposal that was never implemented was the electrification of the Croesor Tramway!

Kellow retired to Cheltenham in 1935 but in 1941, wary of enemy air-raids, he returned with his wife Martha and Alice Mahon, his nurse, to stay for nearly two years at Cartrefle, the Old Post Office, in the High Street at Penrhyndeudraeth.

At one time it was the custom for the Editor of the *Quarry Managers' Journal* to approach senior retired quarry managers and ask for some autobiographical notes. In most cases the result was a few brief paragraphs. Moses Kellow was different! His autobiography runs to nearly 78,000 words.

By 1941 Kellow was almost completely blind. Because of this he had to dictate every word of his 'few brief paragraphs' to John Griffiths. "Griffiths Bach" worked full-time in the Minffordd office of Evan Morris (Morris & Co. Slate Merchants), so dictation took place during evenings and at weekends. John would go home with the result of a dictation session and return a few days later with the typescript. It was very rare for any alterations to be made, because Moses' head was already full of what he wanted to say in the next section. He had no written sources to remind him. Everything came straight from his memory. The result appeared in 24 monthly issues of the *QMJ* during 1944 and 1945. The work has now been published in book form. [2]

Kellow completed his general education in about 1877 and then received his early training in mechanical and mining engineering under the direction of his father. By the age of 20, around 1882, he was managing Parc Quarry and twelve years later took over the management of the Croesor Quarry. He held these joint appointments until his retirement in 1935. During his long career, he managed the Kellow Rock-Drill Syndicate, formed to manufacture and exploit his rock-drill, for which patents were secured practically all over the world. In addition, he practiced as a consulting engineer and devoted much of his time to research in connection with the development of a high-voltage dc generator but ill health caused him to abandon his efforts. Another interest of his was the production of plastics, largely composed of slate dust, samples of which passed the severest mechanical and electrical tests. He also contributed papers to technical societies, dealing with his professional work.

2 Kellow, 2015

Hugh Beaver Roberts (1820 - 1903) [3]

Hugh Beaver Roberts, the second son of Hugh Roberts, was born in the village of Cilcain, near Mold, Flintshire. Educated at Rugby School, he quickly became a successful solicitor and in 1849 took over the practice of John Hughes in Bangor on Hughes' death. He also succeeded him as Clerk to the Bangor Magistrates, Diocesan Registrar, Chapter Clerk and agent to the Bishop. He married Harriet Wyatt of Llandygai in June 1848. The wedding was extensively reported in The *North Wales Chronicle and Advertiser*[4] as an '…auspicious event…celebrated in this city in a manner highly complimentary to the parties…'. They lived at Bryn Menai, near 'The Look Out' in Upper Bangor, where their five children were born. In 1858, he purchased the Plas Madoc estate near Llanrwst. Having moved to Llanrwst, he and his family were back in Bangor by 1861 and living in Wellfield House near the centre of Bangor. By that time, Hugh and Harriet had three sons and two daughters. Ten years later, in 1871, at the time of the Census, Roberts was living with his wife, a son and two daughters in Leamington Spa, Warwickshire and listed

Plas Madoc, Llanddoged, near Llanrwst.

H.B. Roberts bought the Plas Madoc Estate in 1858.

The building was demolished in 1952.

Coflein

as a Registrar of (the) Court of Probate at Bangor and landowner. They seem to have lived in some style, as the household included a cook, an upper and an under house maid, a kitchen maid and a footman. Roberts' coach man lived with his wife and family in the adjoining coach house. The family were still living there in 1875 according to a notice in *The London Gazette* relating to a patent in Roberts' name for the invention of "improvements in the permanent way of railways". [5] Despite Roberts' ownership of Plas Madoc, his 'residence' is shown as Bangor in the Electoral Register for the parish of Llanddoget. [6]

Roberts combined his legal work as a solicitor with business, speculating mainly in slate quarrying and the narrow gauge railways laid down to transport their products to the ports.

---

3 Includes material from 'The Wyatts of Lime Grove, Llandegai', Trans,Caernas. Hist. Soc. Quoted in Boyd. Volume 1, NG Rails in South Caernarvonshire
4 13th June 1848
5 *THE LONDON GAZETTE*, March 26, 1875. page 1871
6 UK, Poll Books and Electoral Registers, 1538-1893/Ancestry

He had married well, as his father in law James Wyatt was, until his retirement, the Penrhyn Estate Agent, following which he was involved with the development of slate quarries. Robert's success in business must have relied much on the close knit character of the Wyatt family. Wyatt had formed the Croesor Slate Company in 1860 and in the same year, Roberts bought the Croesor Estate and leased it to the Company.

Eventually, Roberts owned the Ffridd Nant Quarry at Penmacho and the Braich Quarry in Upper Llandwrog; he was a director of the Bowydd and Maenofferen Quarries in Blaenau Ffestiniog, the Croesor Fawr Quarry in the Vale of Croesor, the Croesor and Portmadoc Railway and the North Wales Narrow Gauge Railways Company. In addition he was involved in the development of the Flint Marsh Colliery and the formation of the Bangor Hotel Company.

His first involvement with railway promotion was in 1853 when, as solicitor to the promoters, he circularized landowners likely to be affected by the proposed but ultimately unsuccessful Conway & Llanrwst Railway. It seems likely that this involvement caused him to meet Hugh Unsworth McKie, who ran the Upper Croesor Slate Quarry Co, and who was the first, but ultimately unsuccessful, NWNGR contractor. However, perhaps Roberts' main claim to fame was his provision of the land, from his ownership of the Croesor Estate, necessary to build a large part of the Croesor Tramway without way leaves. He arranged for the building of a Spooner designed incline and tramway to connect the Rhosydd Slate Co's operations with the Croesor and thus controlled and benefited from the carriage charges arising from their use.

Earlier he had prepared the lease for the Gorseddau Tramway, which would have brought him into contact with George Augustus Huddart (1822 – 1885), a wealthy local estate owner, a director both of the Festiniog Railway and of the Rhosydd Slate Quarry Co. Ltd at various times. He owned the land used as the terminus on the western arm of the Gorseddau Tramway and lent his name to the eponymous crossing of the Cambrian Railways line by the Gorseddau Tramway.

In 1872, having failed to persuade the FR to purchase, operate or amalgamate with the Croesor and Portmadoc Railway (which evolved from the Croesor Tramway in 1865) Roberts offered to upgrade the relevant section of his tramway for incorporation into the southern end of the NWNGR's 'grand plan'. It seems possible that Roberts persuaded Spooner to incorporate into the Moel Tryfan Undertaking the line from Tryfan Junction to Bryngwyn for, as the owner of New Braich Quarry, he had a commercial interest in getting his slate by rail at least to Dinas Junction.

Besides his railway, quarrying and legal interests, Roberts was a Justice of the Peace, District Registrar at Bangor to the Court of Probate [7] and variously Deputy Lieutenant for Merioneth, Member of the Bangor Board of Health and political agent to the Hon. George Sholto Douglas-Pennant during the Parliamentary campaign of 1868. In 1850 he had been "Perpetual Commissioner for taking the acknowledgements of deeds to be executed by married women" for the County of Carnarvon [8]. In 1901 he became a Director of the newly formed Portmadoc Croesor & Beddgelert Tram Railway but later served only as Company Secretary.

Roberts domestic fortunes had changed by 1881 and in that year he was a 'Boarder' at 65 George Street, Marylebone, London age 60, married but with no employment described

7 The Times 13 January 1858 p5
8 The Times 4 December 1850, p. 8

in the Census return of that year. A letter from him to Charles Spooner in November 1882, concerning an extension to the Croesor & Portmadoc Railway to Beddgelert shows him as still living in George Street and he continued to do so until at least 1894; in 1895 he was at nearby Wigmore Street with a 'qualifying address' to vote at Queen Anne's Gate, which may have been offices that he owned. Roberts next appears in the 1901 Census as lodging at 3 Greenbank, Bangor, age 80 and described as a solicitor and widower. He died there on 23rd April 1903, with probate granted to the youngest of his three sons, Arthur Llewellyn Wynne Roberts, on 17th June 1903. Roberts estate amounted to £471 15s 3d, equivalent to around £300,000 in relative income terms, quite a modest sum given the wide ranging nature of his career. Harriett appears to have died in the mid to late 1890s.

Charles Warren Roberts (1852-1897) the middle son, was a railway engineer and latterly manager of Llechwedd quarry in 1887 [9] and Arthur Llewellyn Wynne Roberts (1855-1919) was for thirty years, secretary of the Royal Literary Fund [10].

Wellfield House, Bangor - the Roberts' family home in the 1860s.

Wellfield House and other Victorian properties on this site were demolished in 1961 to make way for the Wellfield Shopping Centre, itself demolished in 2005 to be replaced by the Menai Shopping Centre.

Bangor Civic Society Robert Caldicott/Peter Brindley

---

9 Gwyn, David, 2016
10 Obituary, North Wales Chronicle and Advertiser, 24th of October, 1919

Charles Easton Spooner (1818-1889) [11]

Charles Easton Spooner was the Secretary and Engineer of the Festiniog Railway Company from 1856 until 1886. He was also Engineer to the North Wales Narrow Gauge Railways Company for a brief period before the railway was first inspected. He was involved in the family company of Spooner & Co. In 1863, he was responsible for the survey of the course and civil engineering features of the Croesor Tramway and was associated with it in various ways for some years after.

Charles was the third son of James Spooner, a surveyor and the engineer largely responsible for the building of the Festiniog Railway and its development over its first 20 years. Born in Maentwrog in 1818, Charles, from the age of 16, along with his oldest brother, James, assisted their father and Thomas Prichard during construction. Charles was therefore trained as a civil engineer, gaining much practical experience that would stand him in good stead in his later life. Charles' training continued under both Joseph Locke and Isambard Kingdom Brunel in the late 1830s and early 1840s. He continued to be involved with the railway under his father, who was Secretary to the Company.

Charles was appointed Treasurer to the Festiniog Railway Company in 1847. In 1856 he succeeded his father as Secretary, sometimes referring to himself as Manager. He held the position for thirty years and dominated Festiniog Railway management and engineering until his health began to fail in 1887.

Charles was faced with the problem of a railway working to maximum capacity yet unable to cope with the volume of traffic on offer. He was also aware that others were seeking alternative routes for the transport of Blaenau Ffestiniog's growing slate traffic. He

---

11 www.festipedia.org.uk/wiki/Charles_Easton_Spooner (Accessed 11/10/2018)

investigated the option of conversion to double track but the added capacity could not have paid for the construction costs involved.

Charles Menzies Holland was therefore engaged to design four small engines to be built by George England & Co. The first two engines were delivered in 1863 and were the first ever successful application of steam traction on a narrow gauge railway.

Later, Spooner commissioned Robert Fairlie to design and build *Little Wonder*, an articulated locomotive ideally suited to a relatively short, heavily curved and steeply graded narrow gauge line. Spooner and Fairlie brought the world to Portmadoc in the February of 1870 for a remarkable series of locomotive trials. These trials and the writings of Spooner and Fairlie influenced the promotion of narrow gauge railways throughout the world. On 27th December 1872, the magazine *Engineering* wrote of Charles, "He shows an earnestness and enthusiasm, we may almost say an absolute devotion for the Festiniog Railway".

In addition to his work on the FR, he formed an engineering company Spooner & Co and was responsible for re-engineering the tram road from Penrhyn Quarry, Bethesda, to Port Penrhyn in 1877 as the Penrhyn Quarry Railway.

Charles married Mary in 1848 and they had five children. Soon after Mary died in 1860 his sister Louisa came to live with him. She was still there at her death in 1886 and Charles died three years later in 1889.

Charles Spooner's Curriculum Vitae

The Caernarfon Record Office holds a copy of Charles' CV in draft in his own hand. In places it is a barely legible document [12]. Although very difficult to read, and in places to understand, there is, courtesy of Dr David Gwyn, a transcription of Charles' handwritten notes providing further insight into his character and achievements. The following is a 'tidied up' version of the CV with some explanatory notes at the end. Question marks indicate illegibility in the original document.

> *'Was educated for the profession of civil Engineer and surveying from the time I was 16 years of age, under with Mr James Spooner, till ? Mr Joseph Locke CE in the north of England and subsequently with Mr I K Brunel in south Wales and have since that time to the present practically continued in the practice of Civil Engineer.*
>
> *The following are some of the constructive works that I have carried out*
>
> *Constructing the Borth Road in North Wales*
>
> *Engineering the construction of Docks at Portmadoc Harbour also and of…various Wharves*
>
> *Design & Engineering of the Rhiwbach Railway with (sic) comprising various self acting inclined planes of peculiar construction and Inclined planes with Stationary Engine power for raising the load,*
>
> *Designing & Engineering ? with Incline Plane at Hendre ddu Quarry, the principal of which Works on the principle of a Catinary (sic) Curve weight and reducing the friction of Rope upon Rollers and advancing the Latter ??? compared with the old*

---

12 Caernarfon Record Office XD97/6487

system and reducing the latter ? rollers as compared with the old system to a Minimum, ??? and reducing strains from Machinery and Brake

Wrysgan Railway & inclined Plane with tunnel on 1 5 curve and 800 feet elevation of constructed on a Catenary curve with inclination varying from 1 in 6 to 1½ -

Also Designing Croesor Railway with various inclined planes /on same

As also the Rhosydd Railway at Quarry /it having a self acting Inclined plane of first elevation and ? on Catinary (sic) Curve varying from 1 in 15 to 1 in 1½ worked with Wire Rope one Drum – also various other in the Mining districts –

Altering Festiniog Railway and rendering it applicable for the use passenger traffic for working the slate and goods traffic in the years of 1862-4 & designing suitable Engines and rolling stock for that purpose,

Designing and Engineering of the North Wales Narrow Gauge Railways

The getting up Mining Surveys for the working of Slate Quarries /designing and selecting Machinery for that purpose / in this and other countries as also for Copper and Lead mining

New mode of fish plate for fastening Rail joints on Railways

Construction of Rail bending Machine securing accuracy of curve for any radius and also for Straightening Rails.'

Notes

'Mr Joseph Locke CE' - Joseph Locke (1805 – 1860) was one of the leading civil engineers of the nineteenth century, and with Brunel and Stephenson formed a triumvirate of great early railway engineers. He was engineer to railways such as the Liverpool and Manchester, Grand Junction, London and Southampton and the Sheffield, Ashton and Manchester.

'Mr I K Brunel' - Isambard Kingdom Brunel (1806 – 1859) a man of many talents and ingenuity and associated with great successes such as the Great Western Railway and failures such as the atmospheric South Devon Railway. Quite apart from railways, his many achievements embraced ships, docks, tunnels and bridges. [13]

Inclines on the Wrysgan Railway and Rhosydd Railway [14] – built on a catenary section, ie the curve assumed by a rope suspended at either end, which impose the least weight on the rollers. The Rhosydd incline is steeper than 45% at the summit.

---

13 Brunel and Locke notes – Simmons amd Biddle, The Oxford Companion to British Railway History, Oxford University Press, 1997
14 Gwyn, David, Welsh Slate, RCAHMW, 2015

Chapter 7

Memories

Background

The Croesor enthusiast, Adrian Barrell, has over the years carried out extensive research into the quarry which in 1991 culminated in the 'The Croesor File', most of which is now held by the National Library of Wales [1]. An important part of that research was an oral history project that gathered together the memories of twelve people, spanning the period from 1920, when slate was still being produced, to 1976, when Cooke's Explosives finally vacated Chwarel Croesor. Three are of particular interest because of their references to the Croesor Tramway and edited extracts are reproduced below. Each memory commences with Adrian's contemporaneous notes of the time.

As Adrian says in his introduction, 'All human memory is fallible, none more than mine, and it is not surprising that these accounts contain some contradictions and anomalies. What is surprising is that the differences are so very few; clearly the air of Cwm Croesor produces retentive and accurate memories! I have not attempted to correct the accounts in any way, for who can say which is correct?'

The late Mrs Sali Kidd of Llanfrothen, February 1996

During the 1920s Mrs. Kidd's father was "Johnny Run", in charge of the horse-drawn trains on the Croesor Tramway.

> "Yes, in the old days Croesor was a really lovely village… someone from every house used to work in the quarry then. There was nowhere else for them to go, really, was there? Not just one from every family, sometimes two or three.

> "I only went there once in my life, to the quarry, to see it working. I remember going up when we were young girls, we were about 12, to see them working. We did go to the tunnel, a few yards inside, but we were terrified in there; it was all wet inside. We saw them with the slates. It was amazing. My father didn't go up to the quarry, only for a very short time, as he was working more on the railway track. He was only there for a few months before he was able to have this job on the Run.

> "That's a very perpendicular run up there (Rhosydd). My father used to work at the bottom of that and another gentleman, he used to take the wagons from there to Blaencwm.

> "From Blaencwm my father used to go all the way through to Porthmadog twice a week with the horses, right down to the quayside in Porthmadog. The horses belonged to Croesor Fawr, the farm, and I remember the names of the horses; there were two of them and I used to go and chat to them.

> "We had many a ride with my father down to the incline. There are two inclines down there before you come to the main road. We didn't go down it, just to the top where they were crewling. (*ie to allow the wagons to run or "crewl"*)."

1 Barrell, 2002

He used to work a lot there. I know he pushed a lot of wagons; I remember he had a bad back once. I think they must have been pushing them along there. Sometimes the wagons used to go off the rails. There was a palaver then ... they had to work long hours getting the rail repaired. In those days, they used to work all week and Saturday up to midday, it was a long week, wasn't it? They didn't have any holidays like they have now, just occasional Bank Holidays.

"After the quarry closed my father used to go up the line for quite a while, to see that everything was all right and because they were clearing some things. don't know for how long, I can't remember now. But he did have a job for some time. The inclines didn't work after the quarry closed, not the bottom ones anyway.

"I remember my father saying that after Mr. Kellow had gone to Cheltenham the chauffeur, Mr. Brindley, was burning a heap of things at Bryn. Kellow had been conducting a choir and his baton was there. My father took it from the fire, he thought it was such a pity, and his walking stick as well. I've got that baton, it's a black ebony one. It's something to remember Mr. Kellow by."

Adrian asked about passengers on the Tramway.

"Well, yes, people in the village, something like that. He never really made it a business. He would give a lift to people, if they wanted a lift up. Instead of walking back with a parcel, he would give them a lift. But then there was a shed by Llys Helen [2], a big shed, and my father used to store things in there and people used to come and collect them. Then he used to go up to the two shops in Croesor. They used to have coal, sugar, tea. Yes, I remember, the village was depending a lot on the Run at that time, to bring things up. Of course there weren't many cars in those days; there was Kellow and Morris Rowlands, and Samuel Jones, who used to keep the Siop Goch ... he had a car.

"My father was working from the Rhosydd and then from the Croesor ... so there were two different Managers. I suppose he was paid by Kellow. I can't remember really. Of course, I was only young so I can't remember where the money came from. Wages were very low. Prices were low too, but when you had a family it was a bit stiff. I remember he was only paid £1.30 a week. That was in the late 1920s. And when he was working overtime he didn't have much more than £2.00 - £2.50. People used to work long hours for nothing, really. They had to hold on to the job they had, to keep the family going, and that's it isn't it? But it's amazing how they managed. Some bought their houses.

"When we were living at Llys Helen we only had half the house, because the other half was an office (for the Kellow Rock Drill Syndicate) and they kept all the books and things there. Davies was his (Kellow's) clerk there, his secretary, "Davies Clerk" we used to call him."

2 "Llys Helen" was the cottage owned by Kellow in Croesor Village and was located close to the Tramway level crossing. The 'shed' stood in the angle between the tramway and the road - to the left of the track and just before the road when facing the quarries.

Mr. Edgar Parry Williams, of Croesor, June and September, 1997

Figure 7.1

Edgar Parry Williams

Mr. Williams was born in 1932 and has lived in Croesor all his life; both his father and his grand-father worked in Croesor Quarry. He is the nephew of the late John Henry Williams who kindly provided me (Adrian) with so much information on the quarry. He spent much of his working life as a shepherd on various farms in the area and has an encyclopaedic knowledge of Cwm Croesor. Latterly he studied at the University College of North Wales at Bangor; then, employed by Meirionnydd District Council but based at Coleg Harlech, he spent two years on an oral history project for the St. Fagan's Museum, recording information from older agricultural workers. Time has now overtaken him and he himself is being recorded; he is only four years older than me so it will be my turn soon! On one of our walks we were accompanied by Mr. Iorwerth Ellis Williams, of Llanfrothen; born in 1926, he was the last turbine driver at Blaencwm and was able to fill in many details. I am very grateful to both gentlemen for giving me so much of their time and their knowledge.

Moses Kellow

"The people of Croesor had a deep respect for Kellow, who was regarded as quite a genius; they had great faith in anything he devised. Kellow's father-in-law, Thomas Williams, was a very religious man and a great benefactor of the area. He was involved in the early development of the quarry and of Croesor village; it was he who built the Chapel. He would have been held in high regard. Kellow stepped into his shoes and was similarly respected.

"One of Kellow's offers mis-fired. The generating plant at Blaencwm was not used at night or at weekends, so Kellow offered a free supply of electricity to the Chapel which, in addition to Sunday services, was the focal point for all events in the village. The matter was put before the elders but, after due consideration, the offer was refused with thanks as they felt that electricity was too much like bringing lightning into the building.

"Kellow did bring electricity to Bryn and to Moelwyn Bank. Bryn was a show-place. Edgar lived there at one time and recalls that the old electrical fittings were still in place, including a large stone slab on which the fuse-boxes had been mounted. Bryn is remembered as a cold house, very expensive to heat and run. In Bryn there is one room on the ground floor which faces the quarry road, giving a good view of who was going up to the quarry; the quarrymen left their bicycles leaning against the nearby wall and it was possible to see who was going to be late for work. In this office there was a little cupboard built into the wall, with a lockable door; it looked like a cupboard for the storage of beverages but this was not the case, for Kellow was teetotal. Potential customers for slate would be brought up by car to Bryn; Croesor slate was rather inferior to the Ffestiniog Old Vein product, so it was necessary to put on something of a show to impress them. Early in the morning of the day when a customer was due Kellow would send a message up to the quarry specifying what slates he wanted. John Henry Williams, Edgar's uncle, would then make the slates, so that they would have a full fresh sheen. These slates would be wrapped in a special square of silk and put in a special leather holdall bag to protect them. Then they would be brought down to Bryn in good time and put into the small cupboard. At the right moment they would be brought out again and shown to the potential customer, probably with the suggestion that they had been there for years."

Edgar was able to enlarge upon his uncle's story of Kellow's turbine at Croesor Fawr. The turbine was installed in a small building between the dairy and a barn. Not only did it drive the churn in the dairy, with the disastrous results already recorded in The Croesor File, it also drove a chaff cutter in the barn. It is said that Kellow attempted to speed up the feed to the chaff cutter by thrusting a broom into the hopper, only to be rewarded with a pile of very short pieces of broom handle.

The Tramway

"Iorwerth remembers that the locomotive used on the Tramway in 1943 was the spare loco from the quarry (*but see Adrian's note at the end*). It was taken down the quarry road on a trailer with two tractors, one in front and one at the rear. At one time the same loco was lowered down the A-C incline inside the quarry with a view to using it between the foot of the main incline and C9. This was not a success; the exhaust fumes were very bad and in any case the loco was too large and became wedged after the first two chambers.

"All the traffic on the tramway was horse-drawn in the up direction. Sali Kidd's father, 'Johnny Run', was responsible for running the trains and also for the maintenance of the inclines. He spent the whole of his working life on the tramway; when the quarry closed he was coming up to pension age and took the opportunity to retire.

"Croesor Fawr farm always had the contract to provide horses for the tramway, usually two big draught mares. When the horses were drawing the empty wagons they were quite capable of pulling something like six to ten wagons, but great care was needed when the horse started to move off. If ten wagons were hitched together a horse could never move them if all the chains were tight. It was necessary to make sure that there was a slack between every wagon, and an extra link was specially provided in each coupling. The horse would be able to move the first wagon; the others would follow one by one as the couplings took up and the horse could then move the entire train. In wet weather the horse would sometimes slip on the wooden sleepers. To give the horse a better grip, stone chippings were put between the rails at the usual stopping places. One such place was in the village, where the train would stop to deliver coal; coal for the school was unloaded directly from wagon to coal-house, and the bricked-up coal-hole may still be seen. There was once a farm worker who was unfamiliar with the special starting procedure and tried to start the train with taut couplings. The horse tried so hard to pull the train that it slipped and went down, injuring itself so badly that it had to be destroyed. Down traffic from Blaencwm along the bottom of the valley was always by gravity, and the rumble of the approaching wagons could be heard from a considerable distance; horses were needed only to return empties, and supplies, to the quarries and Blaencwm.

"The tramway bridge over the Afon Croesor at the foot of the incline down from Croesor Quarry is worth close inspection. Despite over 130 years of battering by the river it remains a perfect rectangle and has a fine slate floor to expedite the passage of the water. It is a credit to the original builders and, as Edgar says, it is a "good example of excellent workmanship, even though completely out of sight". At one time the bridge was the boundary between two farms. During 1949/50, Edgar was working there as a shepherd and had problems with sheep trespassing through the bridge. Wire netting was regularly swept away by floods so he made a swinging hurdle, hinged on an old piece of tramway rail. Similar problems in mountain walls were overcome by making a slate grille which would exclude sheep but allow water to pass through.

"The present gate across the track at Blaencwm power-house goes back to the days when the tramway was working; the impact of many wagons has imparted a distinct curve!

"The village was very well served by the tramway. Edgar's grand-mother married in about 1904. She bought furniture in Porthmadog which was delivered by tramway; Edgar still has the receipts.

"The last official traffic on the tramway was supplies of oil to Rhosydd for their pumps. Deliveries took place every two or three weeks. The oil was brought by lorry in drums to Brynhyfryd; from here it was horse-drawn on the tramway as far as Blaencwm and then crewled up the two inclines. Edgar can remember hearing a screeching noise, looking up, and seeing wagons going up the Rhosydd incline. Slate waste used as ballast for these workings was piled at the foot of Blaencwm incline, but it was often removed by tramway by local people for walls or buildings; there were often good cut ends which made it easy to use."

Writing in 1997, Adrian comments as follows on the locomotives at Croesor Quarry:

"Of the three diesel locomotives, all Rustons, known to have been at Croesor Quarry during the explosive-storage period, only one would have been light enough to work on the tramway between Croesor village and Blaencwm during 1943. This was a 20 hp Ruston (Works No. 211625 of 1941). Unfortunately, in 1943 it was still working in a saw mill in Devon, where it remained until at least August, 1944. The only locomotive recorded at Croesor Quarry in 1943 was the 40 h.p. Ruston (Works No. 198297 of 1939) which was the resident locomotive throughout the explosive-storage period and which rested at Gloddfa Ganol museum until tourist activities there ceased when the quarry was sold in 1997. This would certainly have been too heavy for the Tramway and in any case it would have been needed at the Quarry. Was there a fourth locomotive? Several sources confirm that the locomotive used on the Tramway was a Ruston. Did it indeed come from the quarry, or was it brought in specially and taken away again afterwards?'"

Figure 7.2

Nearly 70 years on, 33/40 hp Ruston number 198297 of 1939, which has moved on from Gloddfa Ganol and is now in the care of the Moseley Railway Trust, sits in the sun on the Apedale Valley Railway in 2010. The slate wagons, perhaps regrettably, are from Dinorwic, not Croesor.

The Moseley Railway Trust.

Repairs to the Blaencwm Pipeline: 1943

In *The Croesor File*, Adrian included some extracts from conversations he had had with the late John Henry Williams, who worked at Croesor from 1921 until the closure in 1930. During one of his visits to John Henry he met his brother, Robert Arthur, who worked at Croesor during World War II. He mentioned that at one time a group of soldiers had worked on the maintenance of the pipe-line from Llyn Cwm-y-foel down to Blaencwm, but he could not remember any details.

In 1995 Ken Hawkes, who had been a member of that group, published privately a small book *The Croesor Valley ... and Memories of a Pipeline*. Hawkes developed a proposal in the 1990s to reinstate the Blaencwm power station, develop it as a museum and to re-open the tramway from Blaencwm to the village. The relevant sections of the book are quoted below. Particularly interesting is the only recorded instance of the use of a locomotive on the Cwm Croesor section of the Croesor Tramway. Note also the use of the Blaen-y-cwm incline for transporting the pipes; at this time the incline was still in use for the carriage of oil fuel to Rhosydd.

> "We were fortunate that the rail link between Croesor and Blaencwm was still operational. Our diesel driven locomotive worked wonders, transporting men and materials without once breaking down all through the summer of 1943. We kept the track maintained; today, sadly, this last section of the original link from Porthmadog has disappeared under a growth of grass, stone and hardcore, but in several places the old sleepers are still visible. If only one could wave a magic wand!

> "For the pipeline repairs of 1943, new pipe was supplied from the Midlands; it was sent by rail to Penrhyn and then by road to Croesor. We then resorted to manual labour, and with the help of our diesel locomotive this was how it was done, first loading the pipes on to rail trucks up to Blaencwm; then came the hard bit of man-handling the pipe lengths up the slopes of the Cnicht, quite a tall order. The gravity system from Blaencwm to the old Winch House (*drum-house*) still existed in 1943 and this was a great help to give a first lift of about 150 feet. The datum at Blaencwm is 500 feet and the new pipe was required at 1,250 feet and 1,400 feet so there was a lot of pulling and shoving to do, but the labour force proved equal to the task and overcame all the difficulties.

> "Good progress was made so that by the end of June all the new pipe lengths were in position. The heavy gang left and then the pipe fitters and welders moved in to complete the work. Obviously a lot of time was spent climbing up and down from Blaencwm to Cwmfoel. It certainly gave the 'city men' something to think about as most of the pipe fitters came from the London area and it took a little while for them to acclimatise.

> "We finished the work by September, leaving just the testing of the total pipeline before final completion. This meant opening the penstock valve at the dam head to fill the line. For this exercise we positioned one man at the dam head, then one at each ridge on the descent, to check for any leaks, a total of five men. At a given signal the penstock was slowly opened and, as the water filled the pipeline, we all waited with bated breath for 'thumbs up' or 'thumbs down'. The upper sections of the line were holding but, further down, pinhole leaks, about six in all, were spouting. In fact we had a call from Porthmadog telling of a sighting of a water fountain on the line, so we turned off the pressure. We managed to plug these holes with wooden plugs and eventually went 'on stream'."

Chapter 8

Exploring What Remains Today

From Porthmadog to Croesor Junction

The WHR trackbed, now owned by the Ffestiniog and Welsh Highland Railways, includes the Croesor Tramway trackbed from Croesor Junction to Pont Gareg-hylldrem. The part of the former tramway beyond the Llanfrothen road crossing is not in railway ownership.

The tramway lines around the Harbour have long since disappeared. The line of entry to the Harbour complex from the High Street now lies under a row of shops with housing developments beyond. It is still possible to discern the extent of the original Oakeley Wharf from the buildings that stand there now, indicating where the junction between Croesor Tramway and Festiniog Railway metals once lay (see Chapter 3).

Looking towards the quarries, the original line of the tramway, and later the Welsh Highland Railway, from the High Street into, and then behind, Madoc Street and on to Snowdon Street has disappeared under a supermarket and a car park. The new route of the Welsh Highland Railway follows a wholly new alignment from the end of the Britannia Bridge where it turns sharply right before the petrol station to run behind houses and the supermarket on one side and the Llyn Bach on the other. It rejoins the original route on the approach to the level crossing over Snowdon Street. From Snowdon Street onwards, the Railway runs along the track bed of the Tramway as far as Croesor Junction.

Just beyond its level crossing over the standard gauge railway it passes the site of the Beddgelert, later Gelert, Sidings complex now occupied by the Welsh Highland Heritage Railway. Beyond the junction at Pen-y-mount it follows the original tramway course over Pont Croesor and on to Croesor Junction where it turns north west towards Nantmor and Beddgelert.

From Croesor Junction to Croesor Village

Between Llanfrothen Road and Croesor Village, the course of the tramway passes largely through private land and access is consequently restricted. Much of the Lower Parc Incline, the intermediate transition and the whole of the Upper Parc Incline lie within the Parc Llanfrothen Estate, acquired by Clough Williams-Ellis, of nearby Plas Brondanw, in 1942. The Estate also encompasses the upper reaches of the Parc Quarry tramway, the whole of the Parc Quarry workings and the track of the Croesor Tramway for some 350 yards (16 chains) above the Upper Parc Winding House - this latter stretch including the long stone-built embankment that maintained a level course towards Croesor Village.

From Croesor Junction to Pont Gareg-hylldrem, the trackbed is relatively level, rising only some 30 feet over this section of just over half a mile. Beyond the site of the level crossing over the Llanfrothen Road, a section of just over a quarter of a mile on the level leads to the foot of the Lower Parc Incline and the site of the junction with the tramway that led to the nearby Park Quarry.

Exploring What Remains Today The Croesor Tramway

Figure 8.1 (left)

A section of the track bed between Croesor Junction and Pont Gareg-hylldrem.

Dave Southern 2009

Figure 8.2 (right)

The Llanfrothen road near the site of the tramway level crossing. The confluence between the Croesor and Maesgwm can be seen bottom left - the line of the trackbed curves to the right to follow the south bank of the Maesgwm.

The goods shed known as Gwernydd stood roughly where we see the clump of lighter-coloured bushes in the middle distance.

J. Keylock collection

Figure 8.3

The foot of the Lower Parc Incline (looking back towards Porthmadog).

Dave Southern 2009

Page 76

The Parc Inclines lifted the track almost 400 feet in two stages, the Lower somewhat longer than the Upper. There was an intermediate passing loop laid as an S-bend between the two sections.

Figure 8.4

Looking down the Lower Parc Incline towards the location of the previous photograph.

Note Gareg-hylldrem in the middle distance on the right with Moel-y-gest visible on the horizon.

Dave Southern 2009

On reaching the top of the Lower Incline, the track curved to the left in front of the winding house, before curving right towards the foot of the Upper Parc Incline (figures 8.6, 8.7 and 8.8).

Figure 8.5

The Lower Parc Incline winding house, or what is left of it, can still be seen.

Dave Southern 2009

The long Upper Parc Incline was dominated by its winding house, a substantial free-standing structure (see Figure 3.16).

Having acquired the Parc Estate, Williams-Ellis carried out a number of developments, amongst which he developed the winding house, or 'Drum House', at the head of the Upper Parc Incline into a Banqueting House for Lady Aberconway. This redeveloped building has also been described as a 'Summer House'.

Figure 8.6

The site of the Parc Incline intermediate loop, looking back towards the Lower Parc winding house which is to be found beyond the right-hand tree in the prominent clump.

Dave Southern 2009

Figure 8.7

Looking the other way from a point close to the top of the Lower Incline. The transition from slate slab to wire and post fencing indicates that Figure 8.6 was taken from a point close to the pole supporting the overhead power line.

Dave Southern 2009

Figure 8.8

Looking back into the transition S-bend from the foot of the Upper Incline. Note the pole carrying the power line seen in Figure 8.7.

Dave Southern 2009

The Croesor Tramway — Exploring What Remains Today

Figure 8.9

Looking from the road to Croesor Village across the Park Quarry rubbish tip towards Beudy-newydd farm on a winter's morning.

Beyond the farm the top section of the Lower Parc Incline can be seen rising towards the small clump of trees wherein the remains of the winding house are to be found.

The S-bend, with the power pole (see Figures 8.7 and 8.8), can be seen tracking towards the left directly above the farm.

Dan Quine 2007

Figure 8.10

The Upper Parc Drum House after conversion by Clough Williams-Ellis. The original winding house structure is clearly visible in this view from the Croesor side of the building.

Adrian Barrell   2014

Figure 8.11

The rebuilt Upper Parc Drum House looking up from the original track of the incline. The semi-circular extension will offer commanding views over the Glaslyn flood plain.

Adrian Barrell 2014

Leaving the Parc Inclines, the tramway traversed a substantial stone-built embankment running more or less directly towards Tre-saethon and Croesor Village (Figure 3.18).

Towards the start of this embankment, an occupation access was provided, crossed by the tramway on a bridge of steel girders. The girders, severely degraded after 70 years, can be seen here in Figure 8.12.

Figure 8.12

The occupation bridge through the embankment between the Upper Parc winding house - to the right - and Tre-saethon.

The edge of this opening is just visible in Figure 3.18.

Mark Hambley 2009

In Figure 8.13, Croesor School, now closed, is seen on the right with, in front, the building where coal was unloaded from the tramway. The shed where goods were stored until collected by their intended recipients stood to this side of that stone building occupying much of the width of today's road. The 'smooth' surface in front of the distant gate marks the clapper bridge over which the tramway crossed the Afon Croesor on its final approach to the village (see Figure 3.19).

Figure 8.13

The level crossing in Croesor Village, looking along the trackbed towards the Parc Inclines.

J. Keylock collection

A walk along the tramway from Croesor village

Figure 8.14

Croesor Village

Above Croesor Village the tramway trackbed is accessible and can be walked to the foot of the Croesor and Rhosydd inclines. Access to the quarries should not be attempted by this route, but rather by the access roads up the right-hand side of the valley as once used by the quarrymen.

Figure 8.14 shows Croesor Village, looking up the main street towards the Chapel. The tramway crossed this road just beyond the power pylon in the centre of the image and before the yellow parking warnings outside the school. The Afon Croesor passes under the road just this side of the power pole. Llys Helen (see Chapter 7 and Appendix 2) lies on the right close to the first of the parked cars. The entrance to the village car park is seen in the right foreground.

Figure 8.15

The clapper bridge at Pont Sion-goch looking towards the quarries.

Note the distinctive shape of Cnicht in the background.

Dave Southern

From the car park, which lies alongside the Afon Croesor, cross a small bridge over the stream to join the trackbed heading towards the quarries. 16 chains beyond the level crossing, the tramway crossed the Afon Croesor by another clapper bridge - Pont Sion-goch. After crossing the bridge, the tramway continues as a footpath for another 11 chains to reach Bryn-hyfryd.

At Bryn-hyfryd cottages the course of the tramway becomes a farm track which also provides access to Blaencwm; the new power line from Blaencwm joins the course of the tramway here, following the route of Moses Kellow's original power line from Blaencwm to Bryn. The route continues along the valley floor to Blaencwm, past the foot of the Pant-mawr/Fron-boeth incline. The incline joined the tramway at 6 miles 60 chains from the zero point, 49 chains beyond Bryn-hyfryd.

Figure 8.16

The final stretch of trackbed towards Bryn-hyfryd - the cottages seen to the right.

Note the tramway fencing made from slate slabs - typical of much of this length of the tramway.

Moses Kellow's house, Bryn, stands in the trees to the right, across the road from the Bryn-hyfryd cottages.

Dave Southern

Figure 8.17

The incline leading to Pant-mawr/Fron-boeth with the tramway trackbed at the bottom of the image.

Two tracks cross the incline - a footpath roughly half-way up and the road to the Croesor Quarry just below the top.

The letters 'A' and 'B' indicate the extent of the upper part of the original Pant-mawr incline, taken out of use when the link to new workings at Fron-boeth was established.

Dave Sallery

Just under half a mile beyond the Pant-mawr/Fron-boeth Incline, the tramway crossed the Afon Croesor a third time just before reaching the Blaencwm power station and the nearby Blaen-y-cwm incline.

Figure 8.18

The final stretch of tramway approaching Blaencwm. The power station and the nearby incline can be seen towards the left of the image and the course of the tramway beyond the Blaen-y-cwm winding house is clearly visible.

To the right of centre, the Rhosydd Incline climbing almost to the skyline is also visible.

Dave Sallery 2006

This incline has a wider formation than is usual; this is said to have been provided for the passage of horses on the relatively gentle slope. The drum-house is set back from the incline and the route of the tramway curves across the front of it rather than passing through. The drum-house is a massive slab construction, now devoid of drum and roof. A landing, with two plummer blocks for the remote brake lever, projects forward from the drum-house to give a good view of the incline when crewling. The water pipe from Llyn Cwm-y-foel to Blaencwm runs down the north side of the incline.

Figure 8.19

The restored bridge over Afon Cwm-y-foel, looking towards the Rhosydd and Croesor incline junction with Bwlch Cwm-orthin beyond.

Dan Quine 2007

The tramway now takes up a shelf-like position on the north side of the valley. Rounding a bluff, the course crosses the Afon Cwm-y-foel; the bridge here had collapsed but has been restored as part of what is now a footpath.

Soon the tramway reaches the end of the valley floor and arrives at the junction with the Croesor and Rhosydd inclines. The ruined building at this point is Beaumont's turbine house.

Figure 8.20

The junction between the Rhosydd and Croesor inclines, looking down from the Croesor link.

Between the junction and the start of the incline, the track to Croesor crosses the Afon Croesor for the final time.

Note the restored Afon Cwm-y-foel bridge to the left of the junction.

Dave Sallery 2006

The Rhosydd incline is approached through two banks of waste; this accumulated as ballast for fuel oil and other supplies taken up to Rhosydd after slate production ceased. See Lewis and Denton's excellent *Rhosydd Slate Quarry* [1]. On the slopes to the south of the incline are several wrecked wagons which took the wrong turning.

From the junction the route to Croesor Quarry passes between two slate gate-posts, crosses the Afon Croesor on a slab-built bridge and immediately ascends the final incline to the quarry.

Unlike the Rhosydd incline, this one is steep but climbable. Half-way up the incline the course crosses the long-established foot-path from Croesor village to Bwlch Rhosydd; an occupation bridge is provided under the incline. Above this point the incline passes through a cutting which figures in one of the cautionary tales in Moses Kellow's autobiography. Here and there a length of rail, missed by the scrap-men, lies buried in the grass; one haulage cable extends the length of the incline, obviously not worth recovering. As with the Rhosydd incline, there are the remains of several wrecked wagons.

1 Lewis and Denton, 1994

Croesor Quarry

In 1993, Adrian Barrell visited the various sites in Cwm Croesor and compiled a detailed description of what he saw. The following section is reproduced largely verbatim from his detailed notes of the time [2]. Much of what he saw and described is still in evidence but readers should remember that his descriptions were written 25 years ago and the weather and the hand of man will in some cases have exacted their toll.

His introductory words of caution should also be noted by anyone tempted to explore. 'Please bear in mind that all of the sites are private property and that where appropriate permission should be sought before a visit. A visit to the upper sites will take you to exposed areas where the usual mountain-walking procedures and precautions apply. Underground exploration should be undertaken only by those suitably equipped and experienced. The following descriptions and accounts are records of what has been seen and done; they are in no way a recommendation that you should do the same.'

Croesor Quarry ... above ground.

Figure 8.21

Croesor Quarry mill level photographed in 2010, 17 years later than Adrian Barrell's description.

The road from Croesor can be seen entering the site to the right of the image. The remains of the winding house are seen just above the centre of the image.

Dave Sallery 2010

The road from Brynhyfryd up to Croesor quarry remains in good order. It passes Bryn, the manager's house, occupied by Thomas Williams in the time of the Croesor United and Croesor New companies, and subsequently the home, office and laboratory of Moses Kellow. The house is still inhabited; on the roof are examples of Parc ridging. The road was the original route to Croesor Quarry. Subsequently, when the Croesor Tramway was opened, the road became a secondary route, used only for access on foot or horse-back (Moses Kellow usually rode up from Bryn). During the Second World War it was restored for vehicular use; the small quarry beside the road near the Pant Mawr incline was used as a source of roadstone. During this period, when the quarry was used

2 Barrell. 2002

as a store for explosives, it was customary for loaded lorries to keep to the inside when passing other vehicles. Near the top of the road a wrecked lorry chassis lies lower down the hillside. This was not a genuine accident; the crash was staged for a film sequence. At least two of the drains under the road have been made from old Blaencwm 12" pipes.

Approaching the quarry, the road passes the ruins of a small rectangular building (12'6" x 11'3"). This was the original magazine or powder house. It is said to have been abandoned either because it was too small or because it was too near the road; the latter is more likely, as the second magazine is only slightly larger.

Round the last bend before the quarry are the remains of a larger building (18' x 33'): this was the stables. Originally it had an upper floor which was used as an extra barracks, customarily by men from Rhyd.

The main barracks was also a two-storey building; it is roofless now, as are all the buildings and the wall height is reduced to single-storey or less. Most of the door and window openings remain and the interior layout is still clear. There are traces of the access to the upper storey, in the rear (south) wall at the west end. The 4' wide path leading along the rear of the barracks, giving access to the external stairs, has now largely collapsed. There is no trace of any staircase, internal or external.

The second powder-house is on the mountainside to the west of the barracks; it is reached through a gap in the wall, along a clear path.

Arriving at the quarry past the ruins of the barracks, the gate to the mill area is by-passed by some wooden steps. Immediately inside the yard, on the right, is the combined office and weighhouse, roofless but otherwise intact, although recent vandalism has reduced the height of the walls. At the far end of this building is a brick-built lavatory with the incongruous remains of a pink water-closet! The water supply for this and other domestic purposes came from a very small reservoir on the hillside above, not to be confused with the larger reservoir near Shaft 1.

Here too is the entrance to the lefel fawr; note the bench-mark outside (1563'). The main entrance to the lefel fawr is blocked but presently there is access at the point where the fan ducting passed through. The fan has long gone and most of the stonework for the housing is at Gloddfa Ganol, but the outline of the fan remains on the wall of the adjacent building. The next building was used as a store for oil, timber and other materials. Near the one embrasured window a 2 ½" pipe protrudes from the wall, purpose unknown. The final section of this range of buildings was the joiners' shop; it now contains a substantial concrete bed on which was mounted a Ruston diesel stand-by generator installed by the Ministry of Works in the early 1940s. The roof of this building has been blown right over the rear wall.

The smithy is adjacent to the path up to Llyn Croesor, on a bank above the main mill level. The framework of the bellows, made by Linley and Bingham of Sheffield, is still in place.

The older mills are set back into the hillside. The eastern section became the 'Keldril Works' in 1907; the western part remained in use as a mill until the 1930 closure. Neither part has been used since then. This older section contains the wheel-pit for the original 28' wheel. In later years this wheel-pit was floored over, and the area was partitioned off to make a saw-sharpening shop. A concrete bed above the wheel-pit may have been installed at this time.

The newer mills, surrounded by a chain-link fence with concrete posts erected by Cooke's, were subsequently used for storage by the Ministry and by Cooke's. The northerly projection from the newer mills was the fitting shop. The pit for the newer 39' water wheel is in this part, under the fitting shop; the pit is now refilling with waste after the 1978 Plas excavation, but there are traces of the sluice-gear for the wheel and it is still possible to pass into the western section the old shaft tunnel which runs under the north side of the mill. The tunnel is approximately 6' high and 3' wide; it is intersected at regular intervals by the slabs upon which the shaft bearing were mounted. The slabs are approx. 3' above the floor in order to provide clearance for the pulleys driving the saws.

The entire range of buildings is roofless and in most places the wall height has been reduced to about six feet; other walls have collapsed entirely. Fallen brick pillars show where Kellow's lineshafting ran. The roofing slates were removed after abandonment and the battens lie in piles by the walls.

To the north of the newer mills the slab-walled tailrace of the wheel, partly filled with rubble, crosses the stack-yard towards the edge of the tip. Apart from this, the stack yard is almost a barren expanse; with the eye of faith it is just possible to pick out the lines of the low dividing walls between which the finished slates were stacked. The small hut on the tip was built by Cooke's; it was used for re-packing explosives. On the tips the course of the tramways is still clearly delineated; elsewhere later usage has obliterated all traces of the routes.

The drum-house for the main incline still stands, reduced now to the two main walls and Kellow's 12" x 7" timber buttresses. The roof and drum have gone; out on the crimp there remains only the pivot for the brake lever. The east wall remains vertical but the west wall has begun to lean ominously. The situation of the drum-house at one end of the tip gives an unusual opportunity to see the massive foundations on which it was built; these are best seen from the road past the barracks. At the south end of the drum-house, the end remote from the crimp, there are hinge pins in the walls; presumably these supported gates which could be closed across the tracks.

Croesor Quarry ... underground.

The lefel fawr is open for its entire length; the total length is 1279' from daylight to the railings protecting the lower incline, at the far end of the pass-by. The first 135' is approximately 16' wide, presumably to accommodate the fan ducting; there are clear traces of a dividing wall. The broadening is to the east; the main line of the level, with the tramway, is on the west side.

Presumably doors at the end of the adit were kept closed to encourage the draught from the fan to continue into the quarry. The height at this end averages 7', which would have brought the overhead wire very close to passing heads.

The level narrows to approximately 8'. There is a drainage channel on the east side, slabcovered for the first few yards and open thereafter. Along the east wall is a line of rusty cablehooks which date back to the original Kellow installation. Both walls carry more recent hooks from the Ministry re-wire in the early 1940s. Here and there may be seen the remains of the supports for the overhead wire. (The over-head wire system in the open air was scrapped in the early 1940s, but was not removed from the lefel fawr until after Cooke's had left.)

At 344' from daylight there is a refuge on the east side, and at 366' is Shaft 1; a small fall has come down the shaft, and a 1" bore pipe protrudes from it. At 626' there is another refuge on the east side, and at 753' Shaft 2 is reached; this is the source of the water in the level, and the floor is usually dry beyond it.

At 793' there is a small chamber on the east side; just beyond it, at 829', is the mesh gateway installed by Cooke's in approx. 1966-7. Beyond here, at 875', there is a second refuge on the east side.

At the inner end the level widens into a marshalling area known as the pass-by. This area is 132' long x 33' wide and accommodated four parallel tracks. At the far end a tunnel curves east; this was the access to Floor A East, now blocked by the 1936 fall. The western equivalent collapsed into B1W in Kellow's time (see his Autobiography); it was replaced by a new access which still exists. The bridge over B1W has gone and the chamber is flooded; beyond it the bridge girders remain across B2W but B3W beyond it is bridgeless and, of course, also flooded.

The water in A1W is contaminated with transformer oil, which makes navigation unpleasant; divers have been down to 27m and found a wagon at the bottom.

The brick hut in the south-west corner of the pass-by is built on the original route to A West; latterly it housed a Ruston diesel stand-by generator installed by the Ministry to provide power for the incline and lighting in the event of a failure. In the pass-by the supports for the o.h.w are more obvious than in the level.

At the south end of the pass-by is an area 66' x 24'. Within this area, protected by railings, is the flooded incline down to Floors B and C. Beyond it are the foundations for the lower incline winding gear, originally the Robey portable, later Kellow's electric winder and finally the Ministry installation. The steel tube of the Robey flue can be seen passing through the wall which partially blocks the old upper incline. A substantial slab platform on this wall was for the driver.

The upper incline may be reached by scaling the wall. On the west side it gave access to old B1W, C1W and D1W by means of bridges which have now been removed. Access to chambers on the eastern side has been impeded by the 1936 and subsequent falls Erected across the top of the incline there is a steel grille; this was installed in 1967 and the date, together with the names of the erectors, is scratched in the concrete at the foot of the grille. Immediately beyond the grille is the foot of Shaft 3.

The flue for the Robey runs the length of the incline; it consists of earthenware pipes approximately 18" dia. x 30" long. There is no trace of the original parallel trwnc inclines; at the top of the incline are a few odd pieces of pit-work and the remains of an iron-framed wooden kibble.

A small and otherwise isolated section of Floor A East is accessible from the western workings of Rhosydd. A roofing shaft, said to have been cut originally for Moses Kellow's survey, descends from Rhosydd 16W to Croesor A. It gives access to Floor A on 18E and 19E, beyond which the bridges are missing and only the very intrepid may penetrate. The layout of the western end of this gallery suggests that Kellow intended to extend his bridge avoiding route for the locomotive thus far.

Croesor Shafts 1-3

Shaft 1 is close to the small reservoir above the entrance to the lefel fawr. The shaft is open, a hole partly covered by turf supported on rusty rails. This is the shaft which contained the water balance used to raise slate from the early sinks.

Shaft 2 is covered by a cairn of stones; the waste tip is nearby.

Shaft 3 is high on the side of the Moelwyn. The shaft is still open. There are some substantial timbers which presumably formed part of the pitwork; one baulk carries a plummer block.

On the ground alongside the shaft are two large sheaves, all that remains of the original mass balanced twin-trwnc system in the upper incline. The height of the remaining wall at the shaft collar, when taken in conjunction with the amount of fallen masonry, suggests that at one time the shaft and sheaves may have been protected by a roof.

The Trial Shaft

The rectangular slab-walled collar of the Trial Shaft is almost covered by later tipping. The Shaft is full of slate waste and, presumably, water. From it descends the inclined plane which carried the balance for the drilling rig. This is marked "old incline" on some maps and has been mistaken by some sources as an earlier tramway access to the quarry. Nearby is the tip which contains the spoil from the sinking of the Shaft; it is more extensive than it appears at first sight. Moses Kellow maintained that this tip contains industrial diamonds worth £50,000 which had come loose from the bits of the drilling rig (he was, however, apparently prone to hyperbole !).

The building which housed the turbine and compressor for the Trial Shaft is by the junction at the foot of the Rhosydd and Croesor inclines. The building was 30'6" x 14'6"; all that remains are the very substantial foundations and the north end wall, which has a rectangular wall at ground level to admit the water pipe. Studding protrudes from the pit in which the turbine was situated and there is a duct by which the tail water left the building.

The turbine was fed from the Afon Cwm-y-foel and the line taken by the pipe can be seen rising up the hillside to meet the clear remains of the leat. Above that point the line of the pipe continues, to terminate at two slab pillars; these supported a wooden launder which was fed by a leat from a small tributary of the Cwm-y-foel. Presumably this earlier supply, although giving a greater head, was found to be inadequate and was replaced by the direct leat from the Cwm-y-foel.

Remarkably, the line taken by the compressed-air pipe from the turbine-house up to the Trial Shaft is still faintly visible.

Llyn Croesor ... water supplies

Llyn Croesor is almost empty and the dam has subsided in the centre, despite a large buttress on the outside. There is no trace of the sluice-gear, and the water now vanishes down a hole in the inner apron of the dam.

The spill-way, still in good condition, is separate from the main dam; it discharged to the east into a channel which passed close to the small Upper Croesor working. From the foot of the main dam there is a line of slab pillars which supported the original wooden launder to the mills. Further on, the course of the launder is cut through small outcrops of rock; the channel is approximately 28" wide, suggesting that the launder had an internal width of 24".

The course continues on a rock shelf above the east end of the mills; it then looped round to the back of the mills in order to enter the older mill from the south. On this final loop the main water supply was augmented by a small spring diverted into a subsidiary leat; this can be seen crossing the path from the quarry to Llyn Croesor. Many of the supporting pillars for this final section of the launder are still in place but are not immediately obvious.

Moses Kellow installed a Pelton wheel before proceeding to electrification. It was fed by a 9" steel pipe. This was supported where necessary by collars mounted on the slab pillars which had carried the launder; the pipe has gone but some of the collars remain. The pipe followed the course of the original launder as far as a point above the east end of the new mills, where it descended through a cleft cut in the rock and crossed a bridge to enter the building. The cleft is shown on the 25" O.S., and the bridge is still in situ.

Ceunant Parc (Parc Gorge)

The over-ground remains of Parc are to be found both above and below the road from Garreg to Croesor. From the Croesor road about half a mile above Tan-lan a track leads down across a field to the Parc mills, which lie in the bottom of the steep-sided Cwm Maesgwm. At the foot of this track is the office, gloriously slate-hung and still with Parcro ridging; it is now in residential use.

Upstream on the north bank is the Upper Mill. This contained a 39' back-shot water-wheel which powered the mill and also wound the short incline from the main mill level; it is said to have also powered a compressor. The wheel has gone but the impressive pit remains. At some time the bearings for this wheel were in a bad state; there are clear marks where the wheel has rubbed the side of the wheel-pit. The wheel was fed by a leat and launder from a weir higher up the Maesgwm, but there is little trace of this system. The mill building beyond the wheel-pit was used for the manufacture of Parcro ridging, and odd broken pieces are still to be found on the site.

On the other side of the river is the run-in entrance to Floor C, the main entrance to the workings. A launder from the tail-race of the upper wheel crossed the river and the adit entrance to feed a second wheel in the Lower Mill on the south bank. No traces of this launder remain, but the shell of the mill and of the adjoining smithy still stands. Several slab bridges cross the river; one of these, perhaps more, carried tramways. There was no stack yard in the ordinary sense as Parc produced mainly slabs; these were stacked just to the east of the office.

Figure 8.22

The office building at Parc Quarry (Ceunant Parc). The workings lay to the right.

The 'ditch' seen on the left channels the Afon Maesgwm through the Ceunant Parc site. The track leading to the rubbish tip seen in Figure 8.9 crossed this channel in front of the building to pass between the building and the river as it headed left.

The line to the inclines leading down to the Croesor tramway ran on the near side of the channel. This route crossed the Maesgwm between the two inclines. (See Figure 3.15)

By permission of Llyfrgell Genedlaethol Cymru/National Library of Wales

West of the office the tramway divided into two branches, to the north and south of the Maesgwm. The northern branch takes up a shelf-like position and runs to a substantial tip which was created when the restricted tipping space to the south of the river was full. The southern branch is less obvious, although it is the main line; it runs along the southern edge of the older tips to reach the drum-house at the top of the Parc No.1 incline.

The two Parc inclines were single-tracked with a passing loop, although in fact the loop extended for most of the length of the incline. As a result the top drum-house is very narrow, only 6' between the two walls. Both substantial walls remain in good order but the roof and drum have gone.

The No.1 incline has been almost obliterated. At the top it was tipped over during the re-working of the tips by Mr.E Williams in the late 1930s; by this time the branch was out of use and the slate was removed by road. At the bottom the formation has been badly eroded by the winter Maesgwm. Just beyond the foot of this incline the tramway crossed to the north of the Maesgwm; only the abutments of the bridge remain. Once across the river the tramway takes up a shelf like formation to the top of the No.2 incline, which has been completely obliterated by a farm road. One wall of the drum-house remains; this drum-house was unusual in that it had an overhead horizontal sheave in place of the usual drum. Braking was either on the rim of the sheave or on a separate drum; memories differ. Beyond the foot of this short incline the branch joins the main line of the Croesor Tramway.

Returning to the road, the apparently run-in level in the field just below the road is remembered by Mr John Williams as a shaft fenced with slate slabs; subsequently it was filled in but the site is still visible. A little further up the road, to the south and behind a sheep-fold, is a small quarry which once gave shaft access to the Parc workings; it was used by the County Council as a dump for peaty waste and is now entirely filled. A little further up the same road a slight bulge in the wall denotes the collar of Twll Mwg; despite the name this was the air shaft from the large chamber at the top of the main incline; it is capped.

Further up the side of the valley, are the various trials of the Rhaiadr Copper Mine. Described in the 1932 Auction Catalogue [3] as follows:

'On the Park Estate, in the Rhaiadr Wood, near the Park Quarry, a Dyke containing deposits of Copper can be traced on the surface for upwards of half-a-mile. Tunnels have been driven into this Dyke near its Western End, and a quantity of Copper Ore has been taken out. Other trials, partaking more of a surface character, have been made further to the East, and the indications are that there are Copper deposits of considerable extent. These Copper deposits can be approached by adit levels from the side of the hill, and it is very probable that the workings of the Park Quarry, with their traffic facilities, could be utilized for their development in depth.'

Beyond these workings is the reservoir for the hydraulic winding engine and Moses Kellow's injector.

The reservoir is 8.4m x 8.2m; the present depth is 1.5m. The valve chamber still exists as does some of the fencing, which is of wire threaded through slate posts. The feeder leat heads to the east to the point where it became a wooden launder across a tributary of the Maesgwm. There is no trace of launder, weir or sluice.

Approximately 150 metres west of the reservoir and at about the same height there is a development level which is not shown on any plan of which the author is aware. It is 136 metre long; the initial heading is 292 deg. magnetic, swinging to 302 deg. magnetic after 54m. It appears to be heading for Hafodty Quarry and is on roughly the same bearing as the trial tunnel planned for Floor C but actually driven on Floor F.

---

3 Croesor Quarry closed in December, 1930. During the early 1930's there were at least three attempts to sell it. The Sale Documents quoted from here ran to 22 foolscap pages.

Chapter 9

Welsh Highland Railway

During 1920 a report was commissioned from a Major G C Spring [1] by the Festiniog Railway on the condition and prospects the NWNGR, the PBSSR including the Croesor Tramway and the FR. He made a number of comments on the tramway including:

The tramway starts from Portmadoc harbour passing over reclaimed land to a flat crossing over the Cambrian Railway on the level, which is controlled, by a signal cabin with trap points on the narrow gauge line either side.

No locomotive had been over the branch for some time with the track being in a state of disrepair. The track comprises of flangeless siding type rail with 20 percent requiring renewal before a shunting locomotive can pass over it i.e. an England type engine. The points and crossings are of short lead tramway type and should be replaced before an engine can pass over them.

There is no telegraph, but a link by telephone line exists from the Park quarry office 5 miles from Portmadoc to the Cambrian Wharf siding.

The Welsh Highland Railway (Light Railway) came into being in 1922 with the granting of the Light Railway Order (LRO) of 30 March 1922 (backdated to 1 January 1922 by the LRO, but which had been applied for in 1914) that aimed to link the PBSSR and NWNGR. The originally requested LRO was delayed by the First World War. Two further LROs provided for improvements to the railway's alignment at Beddgelert, a new station site in Portmadoc and a link to the Festiniog Railway. Sir Robert McAlpine & Sons were contracted to refurbish the existing lines and to complete the link between Rhyd Ddu and Croesor Junction, to create a railway that ran from Dinas to Portmadoc.

Figure 9.1

Croesor Junction looking north. The WHR made a left turn just beyond the end of the loop crossover and the Croesor Tramway carried straight on.

There were wagons, presumably awaiting collection, standing on the tramway visible in the centre of the image

C.L. Mowat

The tramway section from Croesor Junction to the bottom incline was in poor condition and some renewal took place using the best of the recovered rail and a few new sleepers after which the track was looked after by the WHR gang.

1 Johnson, 2014

When the WHR took over the lower part of the Croesor tramway a number of halts, stations and goods sidings were established.

Croesor Junction opened on 1st June 1923 and was 17 miles and 77 chains from Dinas. It became an unadvertised request stopping place by August 1923. A memo from 'The General Manager, Portmadoc' dated 3rd August 1923 reads [2]:

> Please note that during this and next month the trains must not stop to pick up or set down passengers at any place not mentioned in the timetables, except at CROESOR JUNCTION (for Tanlan).

It later appeared in the public timetable on the 26th September 1927. Initially, this was a simple junction with the horse worked tramway but later a 142 yard long loop was created. The body from an FR 4-wheel Quarrymen's coach was installed as a shelter for waiting passengers, and for the telephone and train staff tickets when they were eventually provided.

The next halt, **Ynysfor for Llanfrothen** 18 miles 27 chains from Dinas was opened on 1st June 1923 was a request stop. Ynysfor had a corrugated iron shelter located on the east side of the track 12 ft x 10 ft with a small office. The halt was located on the south side of the level crossing to Ynysfor house with a name board on wooden posts and a bench. In Croesor Tramway days a siding was located on the south side of the crossing facing Portmadoc and in 1924 this was lengthened to 40 feet and trap points fitted. A former FR 4-wheel coach body was situated on the northern side of the crossing and on the west side of the track. The halt was mainly used for the transfer of milk churns to Portmadoc.

Pont Croesor for Prenteg at 19 miles 27 chains from Dinas opened on 1st June 1923 as a request stop and was on a gradient of 1 in 655. It was at this point the railway crossed the Afon Glaslyn and in WHR days a water tower was located at the Croesor Junction end of the bridge. The tank was a former Festiniog Railway brine tank wagon minus its wheels, placed on a pile of sleepers at the northern side of the river bridge. It had previously been used to carry sea water to the swimming pool at Plas below Tan-y-Bwlch where there was a slate sea water header tank. The tank can still be seen at Tank curve to this day. The water tank ceased to be used after Portmadoc New Station came into use but came back into regular use when WHR passenger trains ceased crossing the GWR from October 1928. The shelter was similar in design to Ynysfor and measured 15 ft x 10 ft. The original Croesor Tramway siding here was lengthened from 100 ft to 300 ft with an added trap point.

Pont Rhyddin loop (also spelled **Portreuddin**) (Sand siding) was originally a loop in Croesor Tramway days but the loop was removed to Croesor junction in 1924 with the west points retained as a sand siding with a trap point fitted. The siding was still in situ in October 1926.

Portmadoc New opened on 1st June 1923, but was simply called Portmadoc until 1928. It was owned by the Festiniog Railway with the WHR as a tenant. It consisted of a corrugated iron building with a wooden refreshment room and was to the south of the GWR main line, which was crossed on the level and was staffed until the end of the

2 *WHH* 37 p. 7

Figure 9.2

Portmadoc New was an altogether busier station with a wide range of facilities, at least in the early WHR days when this was the Portmadoc terminus for FR trains and the interchange between the FR and WHR.

Lillywhite Postcard
WHRHG collection
WHR 71

summer of 1928. In 1929, economies meant that WHR passenger services were curtailed to the north platform mentioned below. For seasons 1929 – 1933, a connecting service was usually provided between here and Harbour station. For seasons 1934 – 1937 this station merely provided street access for the few WHR trains which still turned back at the north platform. There were no goods facilities but a passing loop with 150 ft platforms either side plus a waiting room and booking office (unstaffed after the 1928 season) was provided and a water tower for the locomotives.

Figure 9.3

Even after the hut was installed, conditions at Portmadoc New (north) were sparse indeed.

Note the standard gauge wagons on the siding from Portmadoc GWR station. The interchange facilities lay about 250 yards up the line towards Croesor Junction.

L&GRP 2470

Portmadoc New station - north (or supplementary) platform located alongside Gelert Siding was opened from 20th May 1929 as an economy measure to avoid WHR passenger trains having to cross the GWR. Station facilities were non-existent (they were provided south of the GWR at Portmadoc New), but in 1935 a waiting room cum shelter was installed. After the crossing ceased to be used, the narrow gauge loop at the Sidings complex allowed locomotives to run round their trains prior to departing northwards. Passengers had to walk across the GWR into Portmadoc, to reach Harbour Station. In its final form, the north platform comprised a gravel waiting area, corrugated iron shelter, name board and two seats. It was never classified as an independent station – always an adjunct to New station south of the GWR crossing.

Page 95

Portmadoc Old was the Festiniog Railway station at Harbour and passengers for through trains between the Festiniog line and New station boarded the train by the goods shed. Trains between the WHR and Harbour would back into Harbour station and the locomotives run round their trains for the journey back north. The station – originally plain Portmadoc – gained the suffix "Old" in 1923, but that was altered to "Harbour" from 1929.

The Welsh Highland closed to passengers after the last trains ran on 26th September 1936 and to freight from 31st May 1937 (this latter being the official closure date).

Cambrian or Croesor Crossing

Figure 9.5

The Cambrian Crossing in 1928, looking over the standard gauge towards Portmadoc New. The signal box that controlled the Crossing sits in the right foreground with the Portmadoc New buildings beyond. The station's water tower is prominent, as is the chimney of the Welsh Slate Co works which was situated just beyond the line of the Gorseddau Tramway.

LPC1663
WHRHG collection
WHHG 25

The history and operation of the mixed-gauge level crossing has been described in detail by Richard Maund in his book *The Chronicles of Croesor Crossing*.[3] A revised version of this book is available to read online on the Welsh Highland Railway Heritage Group website. A replica of the Cambrian Crossing Box has been erected at Pen-y-mount on the Welsh Highland Heritage Railway and replica gates exist where the modern WHR crosses the Cambrian (now Network Rail) line.

Figure 9.6

The replica Cambrian Crossing signal box installed at Pen-y-mount, looking back towards Gelert's Farm with WHHR metals to the right and the F&WHR line to Caernarfon on the left beyond the fence.

Peter Liddell May 2018

3 Maund, 2009

Chapter 10

The Tramway Today

The Welsh Highland Heritage Railway

In 1961 the Welsh Highland Society was formed with a view to reopening the Welsh Highland Railway. In 1964, this group became The Welsh Highland Light Railway (1964) Co. Ltd. Construction of their line following the standard-gauge branch from Portmadoc station started in 1973 following the acquisition of the site of the Gelert (or Beddgelert) Siding from British Railways in 1971. A substantial works and engineering facility has been constructed on the site of the former farm that was situated in the triangle of land between the Beddgelert Siding and the Cambrian Coast Railway. A museum has also been established.

A running line of about three quarters of a mile has been built from the Porthmadog end of the site up to a replica WHR halt near Pen-y-mount Farm and a passenger service started on 2nd August 1980 along the full length of the Gelert Siding site. From 1989, the halt established at Gelert's Farm enabled visitors to view the new museum as part of their trip along the line (there is no public access other than by rail).

Figure 10.1

The WHHR operates a number of items of original Welsh Highland stock, including the Hunslet 2-6-2T locomotive *Russell*, photographed here on the occasion of its return to service after refurbishment in 2014, the Buffet Car and the Observation, or Gladstone, Car. The Gladstone Car, originally built in 1891, is just visible towards the right-hand edge of the image.

Peter Liddell 2014

The company started trackbed clearance in 2000 to facilitate surveying work towards Pont Croesor. After delays due to, amongst other factors, the 2001 Foot and Mouth Disease outbreak, track-laying was completed as far as Traeth Mawr, roughly half-way between Pen-y-mount and Pont Croesor. Passenger trains were operated up to a run round loop installed at the end of this first part of the extension, from 1st April 2007 until the final working on 2nd November 2008.

In 2009 the railway adopted the title Welsh Highland Heritage Railway.

The rebuilding of the main line south from Rhyd Ddu (see below) made connection with the extension to Traeth Mawr at the end of August 2008. Operations over the line from Tremadog Road, Porthmadog to the terminus at Pen-y-mount continue to this day.

Ffestiniog and Welsh Highland Railways

Construction of a new narrow-gauge railway over the course of the original Welsh Highland Railway commenced in Caernarfon, in the first instance using the course of the standard-gauge line from Caernarfon to Afon Wen to reach Dinas. Construction was organised as a series of phases: Caernarfon to Dinas, Dinas to Waunfawr, Waunfawr to Rhyd Ddu and, finally, Rhyd Ddu to Porthmadog. Only latterly during this fourth phase of construction did the line reach the original route of the Croesor Tramway at Croesor Junction.

Phase 1 was completed in October 1997, Phase 2 in August 2000 and Phase 3 to Rhyd Ddu, the original terminus of the NWNGR, was finally opened in August 2003.

Figure 10.2

The 'new' Welsh Highland has been designed around much longer trains than were ever seen in the 1920s and 30s. Currently the main locomotives in use are from the NGG16 class of Garratt locomotives built by Beyer-Peacock for South African Railways between the 1930s and the 1950s.

Here no. 138 (from 1958) runs around its train at Rhyd Ddu.

David Allan April 2008

Phase 4 of the rebuilding of the WHR from Rhyd Ddu through Beddgelert and Aberglaslyn to Porthmadog was completed in stages. Services to Beddgelert commenced in 2009 and later that same year were extended to Hafod-y-llyn. Services were further extended to Pont Croesor in 2010.

From Pont Croesor, the required work included the re-establishment of the Cross-Town Link, albeit on a somewhat different alignment to the original Croesor/Welsh Highland route, and involved the reinstatement of the level crossing over the standard gauge near the site of the original Portmadoc New station. The new crossing was installed on 1st November 2006 but the complete Link was not finished until 2010. Public services over the full length of the new railway commenced early in 2011.

With 12 miles of track, the Phase 4 extension was almost as long the section from Caernarfon to Rhyd Ddu and included significant engineering features, such as the street running across Britannia Bridge to link with the FR at Harbour Station, four tunnels, four large river bridges and the crossing of the standard gauge Cambrian coast line on the level.

Between the WHHR and the F&WHR, a significant proportion of Roberts' 1862 tramway initiative remains in use today, over 150 years after his ideas were first formulated.

A specimen Way Bill Summary Sheet - See Chapter 4 for details.

Cro.Ty.1898

Croesor Quarry Way Bills for Croesor Tramway : Mar-Aug1898

(NLW:Papurai Bob Owen Croesor, Bocs41.6)

Start of quarry month.

16.03.98		43	14x7m1500	Ono 2141 Croesor Siding stock.
		40	16x8m1100	
		54	16x8m1260	Ono 2131 N.Wales Slate Co.'s Wharf
		12	14x7m1330	
		14	14x7m1430	
		55	14x7m1530	Ono 2141 Croesor Siding stock.
		21	16x8m1200	Ono 2131 N.Wales Slate Co.'s Wharf
17.03.98		51	14x7m1390	
		45	14x7m1300	Ono 2141
		3	20x10m1060	Ono 2087 Croesor Siding stock.
		1	16x8m1160	Ono 2131 N.Wales Slate Co.'s Wharf
		20	14x7m1490	
		48	14x7m1460	
		13	14x7m1430	
		2	14x7m1400	Ono 2141 Croesor Siding stock.
18.03.98		10	14x7m1490	
		52	14x7m1460	
		6	14x7m1290	
		42	14x7m1690	
		21	14x7m1430	
		15	14x7m1400	Ono 2141
		56	18x9m1030	
		41	18x9m990	Ono 2087 Croesor Siding stock.
21.03.98	a.m.	19	18x9m1000	
		40	18x9m1000	Ono 2087
		53	10x8m1160	Ono 2141 Croesor Siding stock.
		4	16x8m1100	
		17	16x8b1360	Ono 2131 N.Wales Slate Co.
		47	18x10s690	Ono 2129 Croesor Siding stock.
21.03.98	p.m.	46	16x8m1260	Ono 2131 N.Wales Slate Co.
		55	20x10m1000	Ono 2087 Croesor Siding stock.
		12	20x10b1060	
		14	20x10b1190	Ono 2011 N.Wales Slate Co.
22.03.98		50	20x10m990	
		54	20x10m1030	
		20	20x10m1000	
		13	20x10m1000	Ono 2087 Croesor Siding stock.
23.03.98	a.m.	56	20x10m1030	
		10	18x9m590	Ono 2087
			16x10b190	
			16x10m300	
		15	16x10m1000	Ono 2130 Croesor Siding stock.
23.03.98	p.m.	48	20x10m1000	
		2	18x9m1000	
		6	18x9m930	
		49	18x9m1130	Ono 2087 Croesor Siding stock.

Appendix 2

Transcription of Moses Kellow's letters.

1881

12.7.81 Robert Davies.

Please send at once ½ ton of your 38/- per cwt. powder in 25lb. barrels. We are completely out.

30.8.81 William Jones

Please send me a box of candles tomorrow by run without fail as I am completely out.

15.9.81 Thomas, Festiniog (Harmonium lessons)

As the irregularity of your visits puts me to such trouble and inconvenience I should prefer having the engagement between us cancelled. I am very sorry to do this but under the circumstances I have no other course open to me as I find it impossible to put up with the continual disappointment.

1.11.81 Thomas, Festiniog (Harmonium lessons)

Please return the balance of money due to me for lessons paid for but not received.

1882

20.1.82 Jones, Tallow Chandler

I expect that your account for candles will be paid some time or other about the middle of February.

27.1.82 Jones, Tallow Chandler

Please get ready for me not later than 10 o'clock tomorrow a box of candles, about eight to ten dozen pounds.

2.2.82 Oldham

It is such slow work getting anything down to Portmadoc by the Croesor line with only one run a day and not always that, and we have so few trucks as well, that it is an instance of "a will without a way". I have borrowed a few trucks to help matters on a little, and I have to go down to Portmadoc every few days to see about getting the empties back.

9.2.82 Gray

I have taken the liberty of borrowing a couple of the Gorseddau wagons for ballast wagons and have lent one of our Park trucks so that at present we may do with one besides. I have one for disposal equal to new which I am prepared to part with cheaper than you could get one made.

1883

28.4.83 Festiniog Rly. Co

The amount of traffic over your line from the above quarry to the South Snowdon, Greaves and Cwmorthin Wharf is 20t 1c and to the Public Wharf 18t 11c. This includes everything from 9 May, 1882, to the present day.

14.7.83 Curwen

Ever since the time that I first wrote to you about the letter I had received from the Croesor Rly. Co. stopping our traffic on the line, we have been unable to send any goods down. A couple of orders we had on hand were called for, but notwithstanding we loaded them up, and brought all pressure to bear upon the men, they were left behind two or three days following. I have been on three occasions to Portmadoc to see Jones the Secretary about it, but was unable to find him at home until last night. I represented to him that we should surely lose the order if their delivery was further delayed and that it was owing to no fault of ours that their a/c had not been paid. He consented to let us send our stuff down until Wednesday but no longer unless a settlement be effected by that time.

2.8.83 S. Jones

You can depend on the weight charged on our invoices being correct. They were weighed on the Croesor Rly. weighing machine which is an exceedingly accurate one and far more reliable that the one on the Cambrian Railway.

I have been informed that the Cambrian Railway cannot be depended on to half a ton on a truckload.

1884

9.2.84 Curwen. Annual Report

The slate trade during the last year has been unquestionably poor and very few quarries have been able to make both ends meet. One or two of the very best at Festiniog have no doubt done so but it has been owing to the exceptional quality of their rock which has enabled them to maintain their production in spite of the ruinously low prices.

25.2.84 Roberts, Stationmaster, Portmadoc

MK has had a complaint from Godson & Dobson about carriage costs on roll ridging.

I really hope that in future you will do your best to prevent these high rates being charged as they are not only driving away our best customers and so causing us considerable loss but will do great harm to the Cambrian Railway, as the customers at once transfer their custom to Bangor, where they all inform me they can get their roll ridges sent from at ordinary slate rates.

20.4.84 Roberts, Portmadoc

Please send by Croesor run tomorrow certain: 25lbs 5in. spikes, 1gall. neatsfoot oil and 2 galls paraffin oil. Two tins will be sent down by run tomorrow to put the oil in -- please send to the run for them.

The neatsfoot oil [1] is to be charged to me and the remainder of the order to the quarry.

1894

3.2.94 Robert Isaacs

I accept your offer for the repair of the saw table provided that you give us credit for the old table face as scrap iron, which I presume is not at variance with the intention of your quotation.

Please advise me when to send it down. I want a thoroughly good job made of it as it is our principal slab table where accurate sawing is important.

The blade sent up is fixed badly out of square and we cannot use it without alteration.

18.4.94 D. Jones

We were very disappointed that no truck was sent up yesterday as we have some urgent orders to send off. Please send 3 or 4 up today without fail.

27.4.94 David Williams

I have been considering the matter of the slate dust and for your guidance I have prepared the following estimate of the cost of the dust delivered to your place in Portmadoc:-

First cost of dust per ton	2s. 6d
Loading into bags and wagons	0s. 6d
Crewling over 2 inclines to the Croesor Railway	1s. 7d
Railway charges	1s. 6d
Hire of wagons	0s. 5d
TOTAL	**6s. 6d**

---

1 Neatsfoot oil is a yellow oil rendered and purified from the shin bones and feet (but not the hooves) of cattle. Used as a conditioning, softening and preservative agent for leather.

20.7.94 R. Isaac

I am sending down the connecting rod of the Fret Saw which has worked slack on the pins at both ends. I want the pin attached to the sliding part entirely covered and a new and longer pin put in the bottom as a crank pin. Please make it a close fit as there is a lot of rattle if there is any slack.

26.7.94 Robert Isaac

Please let me have the Hydraulic Engine and also the Air Engine up as soon as possible. I have some gentlemen coming to the quarry this afternoon who can only stay one night, and I particularly want to get all in working order before they come.

30.8.94 Cambrian Rly., Portmadoc

Two emery discs were sent for from London yesterday, consigned to us to Portmadoc, thence for Croesor Railway to Park Quarries. We are in urgent need of them, therefore kindly deliver them to bearer to come up today if arrived, otherwise send them per Croesor run tomorrow.

29.10.94 D. Jones

I am much disappointed at not getting trucks up today. All our trucks must come tomorrow without fail as we have stuff to send for vessels which will sail this spring tide.

26.12.94 R. Isaacs

I am sending down by the run the cylinder of our Air Compressor. You will observe one of the valve spindles has broken. Please repair it at once, also grind the valve a perfect fit to its seat before sending it up.

1900

30.6.00 M.M. Morris, Glaslyn Slate Works

We thank you for your favour to hand enclosing cheque. The marginal figures and the written amount do not agree and we therefore return the cheque for your kind correction.

We may perhaps be allowed to mention that neither of the amounts represents the sum payable to cover the Account.

3.7.00 Robert Isaacs

We are in very urgent need of the slate wagons and I shall feel obliged if you will deliver the balance at once. I saw one of the recently made wagons with wooden brake blocks which are absolutely useless in wet weather and I cannot understand why they are put on after I have repeatedly given instructions that cast iron blocks are to be used. Moreover, the links connecting to the brake blocks should have several holes drilled in them to enable them to be adjusted to the best position for braking. At present they can only be fixed in a position that gives very little braking power. The sketches will illustrate my meaning. In the first of these the braking power is weak but in the second it is powerful. If you drill a number of holes as I have marked the position of the links can be altered so as to give the required braking power.

2.8.00 J.E. Jones

I have been informed that Joseph Sheffield, Tynygro, Prenteg, Cadwallader Williams, London Terrace, Prenteg, and Robert Jones, London Terrace, Prenteg, together with two men from Portmadoc whose names my informant did not know, have on the last two Wednesday nights taken one of our trucks along your line to go home from Rhosydd Quarry.

Cannot some action be taken by us to put a stop to this constant unauthorized use of our property?

10.8.00 G. Morris

We are in hope that crewling on the upper I have been informed that Joseph Sheffield, Tynygro, Prenteg, Cadwallader Williams, London Terrace, Prenteg, and Robert Jones, London Terrace, Prenteg, together with two men from Portmadoc whose names my informant did not know, have on the last two Wednesday nights taken one of our trucks along your line to go home from Rhosydd Quarry.

Cannot some action be taken by us to put a stop to this constant unauthorized use of our property?

Incline at Croesor Quarry will be resumed this afternoon, and that your slates will therefore be brought down without much more delay

17.8.00 Owen, Isaac and Owen

I find the last new wagon sent out has no brakes fitted to it whereas it was arranged that they should all be fitted with brakes on the four wheels. The Croesor Railway people are constantly complaining of the insecurity and worthlessness of our brakes.

Please deliver the remainder of the wagons on order without delay as they are most urgently wanted.

28.8.00 J. Williams

I was very much surprised to hear from you last Friday that you have none of the wagon bodies yet made. Surely there is no need to await the delivery of the wheels before proceeding with this part of the work?

28.8.00 Robert Isaac

The progress you are making with the delivery of the Slate Wagons on order is so slow that we are put to extreme inconvenience and loss. In fact we cannot get on at all as things are at the moment. If you cannot do it at a very much quicker rate than you have been doing it I must make other arrangements.

24.10.00 Cambrian Railways

I wish to put up a sign with the Company's name and address upon it by the side of the road by the crossing at Portmadoc Station. Will you kindly grant permission to put it up on your ground. In the position I contemplate it will interfere with nothing.

Our slate traffic with you is now a considerable one and is likely to develop. Under the circumstances perhaps you may see your way to grant the permission free of charge.

26.10.00 Hadfield's Steel Foundry

Request for a quotation for wheels and axles, giving dimensions.

22.11.00 Thos. Ward

The rails ordered September 4th, although they arrived at Portmadoc some time ago, owing to great pressure of traffic on the Croesor Railway have only just begun to arrive at the quarry.

Our head platelayer has just been here to complain about them and I think with very just cause.

Rubbings have been taken from the fourteen that have arrived at the quarry and I enclose them herewith. I would not complain about a lot of rails if they varied, say, one eighth of an inch in depth, but to have some as low as 2 ¼" and others illegible.

There is not only the matter of the depth of the rails but the form also. How are such varying sections to be fishplated together.

What is to be done in this matter?

5.12.00 C.S.Davies

We have had a great deal of trouble and most serious complaints from customers as to our slate loaded at Beddgelert Siding, Portmadoc, being knocked all about the trucks and being broken to a fearful extent.

This is due to the fact that the siding is used by a large number of dead buffer trucks which entirely neutralize the comparatively few spring buffer in the train during shunting operations.

I have seen your representative Mr. Boulton and he agrees with me that the only remedy for the evil is to construct an additional short line of railway to receive the dead buffer trucks and to enable the slate wagons to be passed out by themselves. The ground is already made up and would be ready for the laying of the rails and sleepers.

There is an alternative plan which would certainly be much better in the long run, but possibly a trifle more expensive to start with, viz., to lower the level of the present railway by the side of the slate quays about three feet, using the stuff so obtained for making the road bed of the additional road by the side of it.

The railway is now level with the bottom of the quays instead of below it as at Minffordd and other places, and the consequence is that the labour of loading the slates is very much increased.

We are now producing upwards of 400 tons per month and I believe the New Rhosydd are doing likewise, so the slate traffic from this siding should be worth making proper arrangements for.

The letter to C.S. Davies of the Cambrian Railways is accompanied by a sketch, it is assumed by Moses Kellow as it is initialled 'MK'. It illustrates the difficulty of deciphering the Letter Books and has been redrawn. (See page 106)

19.12.00 J.E. Jones

Another instance has just occurred of wrong delivery of goods. A quantity of stay bolts we required for the boiler at Croesor was asked to be consigned as follows: M. Kellow, Portmadoc, thence by Croesor Rly. to Croesor Quarry.

A portion of these have turned up at Park and the remainder are now at the foot of our Park Quarry incline having, I understand, been unloaded from Croesor wagons by your man John surname illegible.

As in previous cases they came up with no label and with no instructions where they were to go. I am now put to the trouble of sending back what has come to Park and the delivery of the goods at Croesor Quarry is unduly delayed.

The Cambrian Railway should carry forward their consigning instructions to you and you should see that you get them.

1903

17.9.03 Adams, Dolgelly

The recent tremendous flood we have had here has washed away a considerable portion of the Ceunant Road leading up to the quarry (Parc) and the Park Farm.

In view of the fact that the Beddgelert Railway scheme [2] is now in course of being carried out and that a station will very probably be constructed at the foot of the inclines, would it not be better to divert the road from the old course . . .

The following undated letter relates to the electric locomotive Moses Kellow had built by Kolben and Company in Prague, for working at Croesor. Witting Eborall were London based electrical contractors and supplied equipment for electrically worked tramways

Undated, possibly September 1903 Witting Eborall

Electric locomotive. When I was in Prague I noticed that the headroom allowed for the driver was insufficient, and that in order to enable an ordinary person to sit erect it was necessary to raise the roof about 4 to 5 inches. I arranged with Messrs. Kolben and Co. to do this, but it was not contemplated that the level of the trolley wire should be varied at all but that the extra space required should be obtained by reducing the space between the trolley wire and the roof. As a matter of fact we cannot arrange our trolley wire to be 6ft. 5in. high without involving a very considerable expenditure in cutting away rock, which I do not wish to incur.

19.9.03 Witting Eborall

Transmission line poles. We have had the holes for these cut for months. The second lot arrived but I could not accept them as they were not according to order. Delivery of the third

---

[2] Possibly referring to the Portmadoc, Beddgelert & South Snowdon Railway project

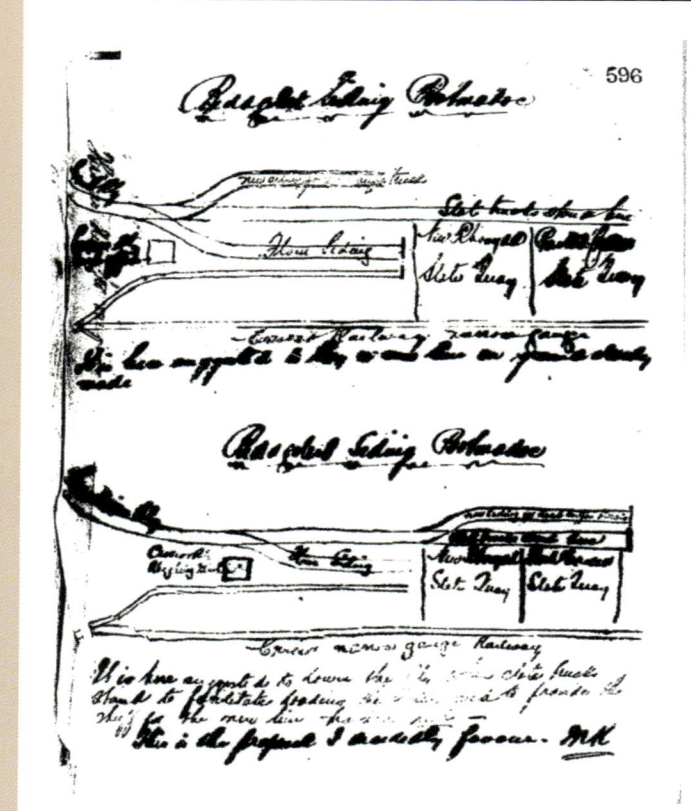

Left

Beddgelert Siding proposal by Moses Kellow - as it survived at the time of Adrian's transcript.

Below

Beddgelert Siding proposal by Moses Kellow redrawn by Adrian Barrell

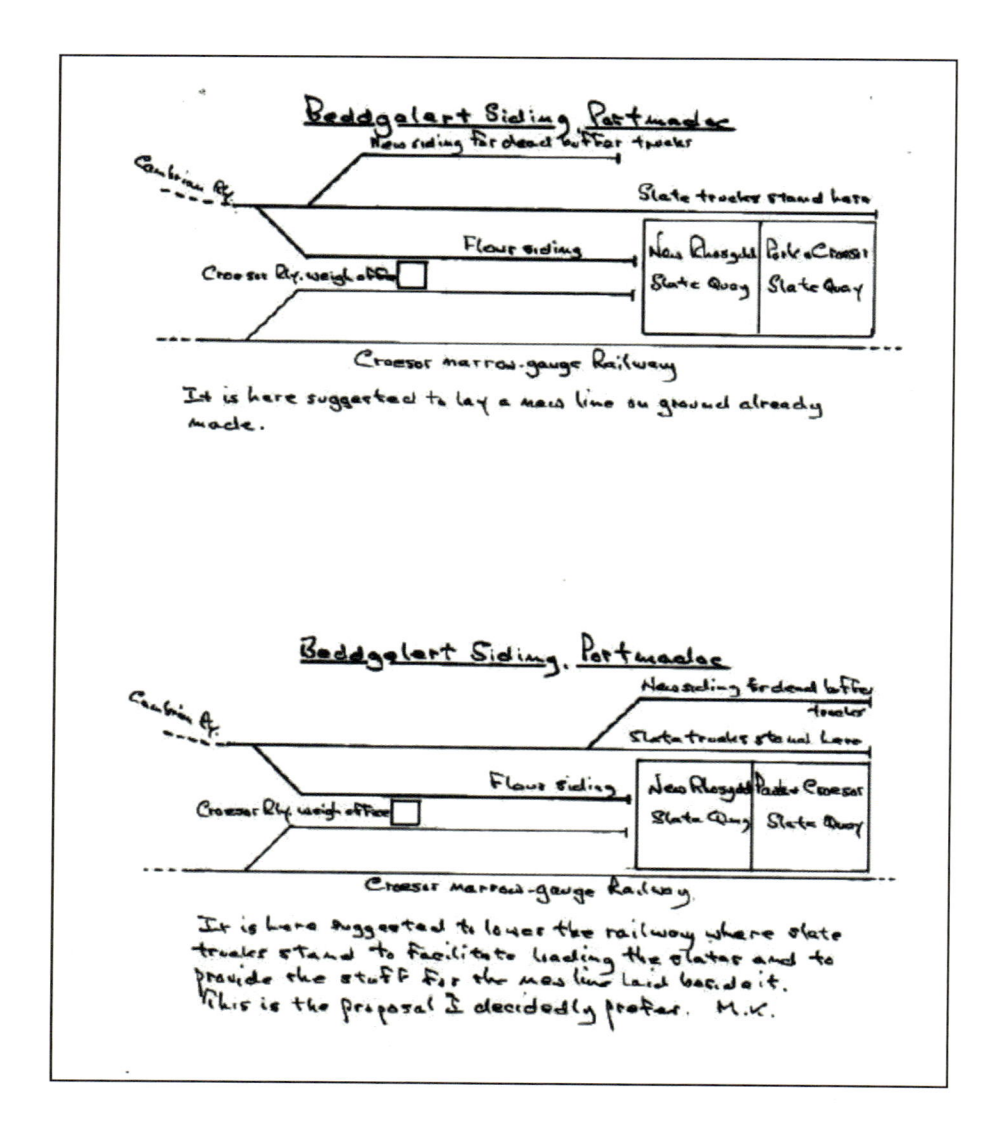

lot is now promised in a few days and we shall proceed to fix them immediately. While the delay with them is annoying if you are anxious to commence erection, I think that there are several things that you might start on practically when you like, the switchboards for instance.

Tests on cable at Glover's Works [3]. I fear it will be very difficult for me to get away next week but I will write as soon as I can fix a date.

Flame Arc Lamps. As I have already told you I have considered lighting one chamber as an experiment.

1904

21.1.04 R. Hughes [4], Rhosydd

The rate hitherto charged by the Parliamentary portion of the Croesor Railway, now the Portmadoc, Beddgelert and South Snowdon Railway, on traffic from Rhosydd Quarry was 10 ½ d. per ton. This formed part of the total rate of 2/7 paid by your Company.

I saw Mr. Aitchison [5], the General Manager, a few days ago and he stated that they are determined to advance this rate to some extent at least. As a result of very strong representations made to him by me he eventually proposed that an advance of 1d. only be made, viz. from 10 ½ d. to 11 ½ d. This would make the total rate including Incline Rent 2/8 as against 2/7. He alleges that they are entitled under their Act to charge 1/9 ½ but before commencing construction are legally able and are willing to enter into an agreement for a lower rate.

I should be glad to know in confidence what you think of this and whether your Company are going to agree to the 1d. advance, as we are in exactly the same position as yourselves. I fenced the matter with him until I knew what you were likely to do.

12.2.04 Morris Jones

Will you please turn two of our loads of coal into Pantmawr Siding, taking the numbers and if possible the weight also, as near as you can get it.

1905

Note the comment referring to ownership and control of the Tramway at the end of this letter to the tax surveyor.

13.9.05 H.B. Simpson (tax surveyor)

Most if not all of the forms relating to Income Tax have been addressed to me at my private house. I prefer it so as regards my own private income but, when the forms relate to the Quarry, Croesor Estate, Croesor Tramway or other business concerns, with which I am connected, I prefer that they be addressed to the office.

You will readily appreciate that business documents arriving at a private house are apt to be mislaid and forgotten, and as I have been absent from home for nearly two months it is possible that something has come to the house which I have not seen.

The Croesor Tramway is part of the Croesor Estate but is under the special control of Mr. Josiah Kellow. I have therefore asked him to prepare the return.

1906

10.1.06 Owen, Isaac & Owen

The saw-sharpening machine we sent down for repair is very urgently wanted and we shall be glad if you will send it up at once.

20.12.06 Thomas Ward

Kindly let me have a definite date when I may expect to get delivery of Rails on order. This delay is causing us great inconvenience.

---

3 Walter T. Glover & Co of Trafford Park, Manchester and London, electrical cable manufacturers, ultimately became part of British Insulated Callender's Cables.
4 Secretary of the New Rhosydd Slate Quarry Co. Ltd
5 Gowrie C Aitchison (1863 – 1928) General Manager of the North Wales Narrow Gauge Railway Company, the Portmadoc, Beddgelert & South Snowdon Railway and the Snowdon Mountain Tramroad & Hotels Company Ltd

1907

12.2.07 Thomas Ward [6]

Two trucks of rail have now arrived at Croesor Quarry ... I find on examining the rails that you have sent them in short lengths of 15 feet which means that there will be more joints than would be the case were they longer, consequently the roads are not so firm.

9.2.07 O.D. Jones

Kindly state what would be your lowest price for building a dry wall with slate building stones, about 2ft 6in thick at the bottom and 1ft thick at the top. It is required as a buttress to the Croesor Railway Cei Mawr.

23.4.07 Thomas Ward

Please quote me your lowest price for 91cwts of 35lb rail in the longest possible lengths.

27.8.07 No name [7]

I am very pleased to hear that you like the cottage. I think you would be very comfortable in it. Butchers, greengrocers and other tradesmen call with their carts regularly for orders, or anything can be ordered from Portmadoc and delivered by the Croesor Narrow Gauge Railway which runs through the valley.

You can hire carriages at the Brondanw Arms Hotel, or if a dog cart would suffice you could get this from the farmer close by. Milk and butter can be got in abundance as there are several farms close by. I think that there would be no difficulty in arranging to get a woman to clean the house etc.

There is a fine view of the Moelwyn Mountains from the dining room window.

Aug., 07 Another letter identifies the cottage as Llys Helen.

20.9.07 Jones

I am arranging for crewlings that will take the Directors who stay at Bryn on the night of Tuesday, 24th, up to Croesor Quarry by 10.30am. I think this will be quite early enough as they will have the whole day before them for the inspection.

Illegible ... Park Quarry inspection the following morning and the Board Meeting in the afternoon.

30.10.07 G.C. Aitchison; P.B.& S.S.Rly

Workmen's Conveyance. The men from Croesor Quarry are running a car once a week under exactly the same conditions as the Rhosydd workmen. I am collecting from them at present 5/10 per month.

The men at Park are also running a little car, just sufficient to accommodate six men. As you know, quarrymen's wages are at present very low and they cannot afford to pay very much for accommodation of this kind, more especially as they have to pay in addition for the pony and car.

I ask you therefore to put the Croesor men's car on the same footing as the Rhosydd men's car, viz. 8/- per month, and if you can possibly see your way to do so to put the Park men's car on 4/- per month as it is only a little car and only six men ride on it.

1908

12.6.08 Hughes

I am returning to you per run today the 2in Yellow Pine plank. I found on examining it that it is full of shakes[8] right through the whole thickness from one face to another, making it quite

---

6 Thos. W. Ward Ltd , Sheffield, Yorkshire, steel, engineering and cement business. The Rail Department supplied light and heavy rails, sleepers, points and crossings and equipped complete sidings
7 This letter, along with a photograph of Llys Helen, was discussed in *Welsh Highland Heritage* Issue 42 (December 2008)
8 Shakes - lengthwise separation of the wood along the grain, usually occurring between or through the rings of annual growth. (blog.spib.org)

useless for us for the kind of work it is intended for, viz. pattern-making. The wood for this work must be perfectly sound

23.7.08 Rother Bros

I have taken this up with out foreman loader at Portmadoc. He denies that the slates were carelessly stowed, and says that not only were all the Best chalked conspicuously but that he took the Captain down and drew his attention to the various parcels.

17.9.08 J.C.Smith

Please supply one dozen bottles of John Davy Special Old Whisky, as supplied to Mr.Thomas Jones of London.

20.10.08 Cambrian Rlys Co

With regard to the case of whisky delivered a short time ago, one bottle value 3/6 was broken. I made a claim on the Croesor Rly Co. but they inform me that the case was damaged when it was handed over by you.

1909

9.2.09 Davies

Beddgelert Railway [9]. I should be please to promote this matter in any way in my power, and as the only member of the Committee on the Merionethshire C.C it will be well that I should be in possession of any information that would help me in securing the attention and sympathy of that body. If you can let me have some further particulars as to the nature of the proposals and generally how the matter stands I shall be glad.

1910

14.5.10 L.& N.W.Rly.; Lost Property Dept

MK left his muffler on a train. It was made of wool, was 7ft (sic) long, dark green with red bands across near the ends. Fringe at both ends about 2in long.

23.7.10 R.J. Low

Park Quarry. The tram lines and incline belong to the Quarry and are included in the Quarry rent. The Royalty per ton is 1s.3d., which merges into a dead rent of £40.

Croesor Quarry. The tramway is the property of the Quarry Company as far as the bottom of the first incline and is included in the Quarry rent.

16.4.10 Pritchard and Davies

We hereby beg to give you notice that you must forthwith carry out the work you have undertaken to in accordance with your contract, which is to clear about 500 tons of material and loading same into wagons on the Croesor Railway. Failing this we shall put other men to do the work, holding you responsible for any loss that may result from the non-fulfilment of your contract.

12.10.10 Josiah Kellow

I am getting up a series of photographs appertaining to the quarry and shall probably use them for publication. I am therefore particularly anxious that the photos are as good as they can be made.

Among other photos I wish to get one of our Wharf in Portmadoc. I want the photo to be taken from the end nearest Portmadoc so as to have the very best quality and the cleanest and straightest piles in the foreground. Anything that in any way mars the view must be put right before the photo is taken, including the clearing of any loose rubbish and particularly of any broken slates that may be lying about the place.

As this photo will probably go before a large number of Slate Merchants I want to make it as impressive as possible and you had therefore better arrange to have, say, two gangs of

---

9 By 1909, the Portmadoc, Beddgelert and South Snowdon Railway projct as on the verge of being abandoned

men loading Cambrian trucks, others unloading Croesor Railway trucks on to the quay, and stick as many men about the place as you most conveniently can.

I will, however, give you this word of warning. They will want to turn round to have their photos taken, whereas the keynote of every photo that I am taking is that the men are vigorously at work, either about their usual avocations, or walking in an energetic manner, and preferably with the tools of their trade in their hands. Don't let them look towards the photographer.

It will require very careful stage management to get the required result and, unless it is got, I will not use the photo and all the time and money will be wasted. I can hardly tell you the difficulties I have encountered here in getting the necessary conditions fulfilled.

1911

16.6.11 Cambrian Railways

We beg to draw your attention to the question of the supply of slate trucks to our siding at Portmadoc station.

We are continually experiencing great difficulty in this direction. We now require several wagons and there is not one available. This condition very often prevails and is fast getting chronic. It is a source of loss to us not only in prestige but also monetarily, as our customers get vexed and tired of waiting and finally obtain their supplies elsewhere.

We must ask you to give immediate attention to this very important matter and take specific measures not only to rectify the present difficulties, which are bad enough, but also to prevent their recurrence and their very irritating consequences.

1912

17.8.12 Williams

Will you kindly send us a copy of the Portmadoc, Beddgelert and South Snowdon Railway account against us for carriage for the month of April. We have the Park quarry part of the Bill but are unable to find the Croesor part.

1913

1.4.13 Alford

We are awfully sorry for the delay, but the traffic along the Croesor Railway has almost been monopolized by other urgent business which rendered it difficult just then to keep up our stock at Portmadoc.

14.6.13 Crick, Festiniog Railway

We are in receipt of yours with reference to Wagon No. 50. We beg to inform you that the wagon was at once withdrawn for the purpose of putting it right for work. We are quite willing to keep our trucks in good running order and beg to thank you for drawing our attention to defective ones.

1914

18.3.14 Queenby (Albert James), Shepherds Bush

Will you kindly walk down Snowdon Street to Messrs. Kellow & Co.'s office. We are arranging for out Manager's brother, Mr. J Kellow, to see you and to conduct you to our Siding Wharf to see our slates in bulk. Perhaps you will kindly ring up the writer on the telephone as we trust some business can be arranged with you this time.

19.5.14 Jones & Co., Portmadoc.

Will you kindly send 2 boxes of candles up to Croesor Quarry per tomorrow's run.

24.11.14 Kellow & Co., Portmadoc

Please forward to Croesor Quarry per tomorrow's run 2 tons of Barracks Coal and 1 ton of coal for Blaencwm Generating Station.

8.12.14 Madison Sterling (*partly illegible*)

I certainly think Leave to Borrow and Authority to Pay the Railway accounts due to the Croesor Railway on 10th July, 1914 amounting to £491.0.2 and to the Portmadoc, Beddgelert and South Snowdon Rly. on 10th July, 1914, amounting to £158.3.0 should be also obtained at once.

The grounds upon which this Authority to pay is asked is:

Croesor Railway.

1. That the Railway is privately owned and that the owner has the right, which he will undoubtedly exercise at an early date, of stopping the traffic from the Croesor Quarry.
2. That there is no alternative way in which the traffic can be sent that is reasonably practicable.
3. That if the traffic is stopped the stock of slates at the Quarry which is worth about £2,800 cannot be sold.

Portmadoc, Beddgelert & South Snowdon Railway.

1. That the amount charged for sending slates to Portmadoc is 11½d. per ton.
2. That we get a rebate on all slates by rail along the Cambrian Railway from Portmadoc of 1s.2d. per ton, and thus the amount of this rebate nearly covers our own indebtedness for carriage.
3. That the rebate is paid by the Cambrian Railway to the Portmadoc, Beddgelert & South Snowdon Railway and that any hitch in the payment of their carriage account would seriously jeopardize the payment of the rebate in the future.

19.3.15 Ellis Griffith

I want to slate a building at Croesor Siding, Portmadoc, immediately on my Patent System and would like you to do the work. I propose going to Portmadoc very early tomorrow morning. I shall be there at 8.0 am to see about it. Can you meet me there in the morning or immediately after noon?

I may say that the Portmadoc job will probably be followed by a very much larger one.

In his autobiography [10], Kellow relates how he was shipping Best Old Vein slates to the Continent, especially to Germany, and that his relations with his 'German customer were of a very cordial character.'

It was usual to book orders for the spring trade in January or even in the December of the previous year but in the early part of 1914, the orders did not materialise, 'notwithstanding the good relations that prevailed between my German customer and myself, nor were any orders for cargoes received during that year at all.'

Up to 70% of Croesor's total output was Old Vein slate suitable for the German market. The lack of orders for this substantial proportion of the quarry's output led to a large build up of stock and a significant slow down in cash flow. The consequence was that the Debenture Interest due on the 1st July, 1914, was not paid.

One of the smallest of the Debenture Holders started a Debenture Holders' action against the company. In order to conserve the assets of the company, it was considered advisable to appoint a Receiver, and Kellow was duly appointed to this role.

The final pages from the Letter Books make clear the additional complications of Receivership.

---

10 Moses Kellow, 2015

Appendix 3

Croesor Quarry Output ('make') and Employment 1895 - 1941

Output and employment statistics (overleaf) for Croesor Quarry during the Kellow period were compiled by Dr. Michael Lewis and subsequently deposited at Plas Tan-y-Bwlch.

We are grateful to have had access to them, via Adrian Barrell.

Fig A3.1

Work in progress on one of the mill floors at the Croesor Quarry in 1911.

Year	'Make' Tons	Men Employed Underground	Surface	Total	Make per man per year
1895		19	20	39	
1896	804	54	36	90	8.9
1897	2194	51	50	101	21.7
1898	2356	67	54	121	19.5
1899	2577	63	59	122	21.1
1900	3750	57	69	126	29.8
1901	3600	79	70	149	24.2
1902	3343	79	80	159	21.0
1903	2926	84	80	164	17.8
1904	3226	70	70	140	23.0
1905	3046	87	67	154	19.8
1906	2742	93	51	144	19.0
1907	2474	75	55	130	19.0
1908	1937	42	47	89	21.8
1909	1957	43	49	92	21.3
1910	2233	40	49	89	25.1
1911	2540	58	49	107	23.7
1912	2856	59	51	110	26.0
1913	2721	46	51	97	28.1
1914		34	38	72	
1915		25	31	56	
1916	ca. 1288	17	29	46	ca. 28.0
1917	ca. 1102	11	27	38	ca. 29.0
1918	ca. 883	12	20	32	ca. 27.6
1919	ca. 707	18	20	38	ca. 18.6
1920	ca. 912	18	30	48	ca. 19.0
1921	915	22	31	53	17.3
1922	ca. 875	27	35	62	ca. 15.7
1923	ca. 1349	42	43	85	ca. 15.9
1924		35	49	84	
1925	1115	39	44	83	13.4
1926	1400	39	50	89	15.7
1927	1349	43	50	93	14.5
1928	1570	39	49	88	17.8
1929	1724	40	50	90	19.2
1930		32	47	79	
1931		0	3	3	
1932		0	4	4	
1933		0	0	0	
1934		2	1	3	
1935		2	1	3	
1936		2	1	3	
1937		2	1	3	
1938		1	1	2	
1939					
1940	ca. 300			ca. 11	
1941	ca. 300			ca. 15	

Bibliography

Extensive use has been made of the following books, magazines, journals and websites in the preparation of this book. The 'footnotes' provide further detail.

Barrell, Adrian, Chwarel Croesor, 'The Croesor File', research from the 1860s onwards, privately published as a CD, Version 3.001, 2002 - a copy of the CD has been deposited at the National Library of Wales as 'File NLW ex 1583. - 'The Croesor File': research material relating to the Croesor Slate Quarry'.

Berks, T, Jones, Dr Gwynfor P, Snowdon Mill: GAT Project No. 2057 Report No. 806, Gwynedd Archaeological Trust, 2009

Booker, N G; 'James Cholmeley Russell - 'An £8 Million mystery', Festiniog Railway Heritage Group Journal, Winter 1996/97

Boyd, J I C, Narrow Gauge Railways in South Caernavonshire, Volumes 1 & 2, The Oakwood Press, 1988 (Vol. 1), 1989 (Vol. 2)

Davidson, A, Jones, A, Jones, G P, Gwyn, D - Gwynedd Quarrying Landscapes Slate Quarries, Gwynedd Archaeological Trust Report No. 129, 1994

Gwyn, David, Welsh Slate: Archaeology and History of an Industry, Royal Commission on the Ancient and Historical Monuments of Wales, 2016

Hatherill, G & A, Slate Quarry Album, RCL Publications, Wales, 2001

Hawkes, Kenneth, The Croesor Valley, Kenneth Hawkes, unknown date

Johnson, Peter, Festiniog Railway: From Slate Railway to Heritage Operation 1921 – 2014, Pen & Sword, 2017

Johnson, Peter: An Illustrated History of the Welsh Highland Railway 2nd Edition, OPC (Ian Allan), 2009

Kellow, M, The Autobiograhy of the Croesor Quarry Manager, Delfryn Publications, 2015

Lewis, M J T & Denton, J H, Rhosydd Slate Quarry, 3rd impression, Adit Publications, 1994

Lindsay, Jean, A History of the North Wales Slate Industry, David and Charles, 1974

Manners J & Bishop M, Ghosts of Aberglaslyn, The Portmadoc, Beddgelert & South Snowdon Railway A History, Welsh Highland Railway Heritage Group, 2016

Maund, Richard, Chronicles of Croesor Crossing, Welsh Highland Railway Heritage Group, 2009
(On-line revised edition available at www.welshhighlandheritage.co.uk/wp-content/uploads/CHRONICLES-revised-website-edition-Apr-2018-1.pdf)

Mitchel, Vic; Keith Smith, Branch Lines Around Portmadoc 1923-46, Middleton Press, 1993

Richards, A J, A Gazeteer of the Welsh Slate Industry, Gwasg Carreg Gwalch, 1991

Richards, A J, The Slate Railways of Wales, Llygad Gwalch Cyf, 2011

Welsh Highland Heritage, Welsh Highland Railway Heritage Group,

In particular:

Welsh Highland Railway Halts, Issue 6, pp 2-3, October 1999

Sale of Railway Lines (Newspaper extract - C&D Herald), Issue 22, p. 3, December 2003

Crossing on the Level, Issue 28, p 3, June 2005

A Welsh Highland Operational Chronology 1922 - 1937, Issue 57, pp 6-9, September 2012

The End of Croesor Crossing, Issue 58, pp. 8-9, December 2012

Working the Croesor Valley Traffic in the 1930s, Issue 58, pp 12-13. December 2012

More on the Croesor Goods Runs, Issue 59, p. 4, March 2013

Chronicles of Beddgelert Siding, Issue 60, pp. 4-6, June 2013

Chronicles of Beddgelert Siding II, Issue 61, pp. 2-4, September 2013

Relaying the Croesor Crossing, Issue 61, p. 7, September 2013

North Wales Narrow Gauge Railways and Welsh Highland Railway (Light Railway): A Station Chronology. Issue 63, pp. 6-10, May 2014

Websites and Internet Resources

britishlistedbuildings.co.uk

wikipedia.org/wiki/Croesor_Quarry

wikipedia.org/wiki/Croesor_Tramway

wikipedia.org/wiki/Rhosydd_Quarry

www.coflein.gov.uk, online database for the National Monuments Record of Wales

www.douglashistory.co.uk/history/families/douglas-pennant

www.festipedia.org.uk/wiki/Cambrian_Crossing

www.festipedia.org.uk/wiki/Charles_Easton_Spooner

www.festipedia.org.uk/wiki/Penrhyn_Station

www.isengard.co.uk/PBSSR.htm

www.llechicymru.info/croesor.english.htm

www.measuringworth.com/uk

www.welshhighlandheritage.co.uk/wp-content/uploads/CHRONICLES-revised-website-edition-Apr-2018-1.pdf (on-line, revised version of Chronicles of Croesor Crossing)

F&WHR Online Photo Archive - http://217.34.233.120:8086/ - for photographs with 'iBase' reference numbers.